British Rail

CO
STOCK

THIRTY-NINTH EDITION
2015

The complete guide to all
Locomotive-Hauled Coaches which
operate on the national railway network

Peter Hall & Robert Pritchard

PLATFORM
5

ISBN 978 1909 431 13 3

© 2014. Platform 5 Publishing Ltd, 52 Broadfield Road, Sheffield, S8 0XJ,
England.

Printed in England by Berforts Information Press, Eynsham, Oxford.

CONTENTS

PROVISION OF INFORMATION

This book has been compiled with care to be as accurate as possible, but in some cases information is not easily available and the publisher cannot be held responsible for any errors or omissions. We would like to thank the companies and individuals which have been co-operative in supplying information to us. The authors of this series of books are always pleased to receive notification from readers of any inaccuracies readers may find in the series, to enhance future editions. Please send comments to:

Robert Pritchard, Platform 5 Publishing Ltd, 52 Broadfield Road, Sheffield, S8 0XJ, England.

e-mail: robert@platform5.com   **Tel:** 0114 255 2625   **Fax:** 0114 255 2471.

This book is updated to information received by 6 October 2014.

UPDATES

This book is updated to the Stock Changes given in **Today's Railways UK 155** (November 2014). Readers are therefore advised to update this book from the official Platform 5 Stock Changes published every month in **Today's Railways UK** magazine, starting with issue 156.

The Platform 5 magazine Today's Railways UK contains news and rolling stock information on the railways of Great Britain and Ireland and is published on the second Monday of every month. For further details of Today's Railways UK, please see the advertisement on the back cover of this book.

Front cover photograph: CrossCountry HST Trailer Composite Kitchen 45005 at Newton Abbot on 05/04/13.   **Robert Pritchard**

BRITAIN'S RAILWAY SYSTEM

INFRASTRUCTURE & OPERATION

Britain's national railway infrastructure is owned by a "not for dividend" company, Network Rail. In September 2014 Network Rail was classified a public sector company, being described by the Government as a "public sector arm's-length body of the Department for Transport".

Many stations and maintenance depots are leased to and operated by Train Operating Companies (TOCs), but some larger stations remain under Network Rail control. The only exception is the infrastructure on the Isle of Wight: Island Line was the only franchise that included the maintenance of the infrastructure as well as the operation of passenger services. As Island Line is now part of the South West Trains franchise, both the infrastructure and trains are operated by South West Trains.

Trains are operated by TOCs over Network Rail, regulated by access agreements between the parties involved. In general, TOCs are responsible for the provision and maintenance of the locomotives, rolling stock and staff necessary for the direct operation of services, whilst Network Rail is responsible for the provision and maintenance of the infrastructure and also for staff to regulate the operation of services.

The Department for Transport is the franchising authority for the national network, with Transport Scotland overseeing the award of the ScotRail franchise. Railway Franchise holders (TOCs) can take commercial risks, although some franchises are known as "management contracts", where ticket revenues pass directly to the DfT. Concessions (such as London Overground) see the operator paid a fee to run the service, usually within tightly specified guidelines. Operators running a Concession would not normally take commercial risks, although there are usually penalties and rewards in the contract.

During 2012 the letting of new franchises was suspended pending a review of the franchise system. The process was restarted in 2013 but it is going to take a number of years to catch-up and several franchises are receiving short-term extensions (or "Direct Awards") in the meantime.

DOMESTIC PASSENGER TRAIN OPERATORS

The large majority of passenger trains are operated by the TOCs on fixed-term franchises. Franchise expiry dates are shown in the list of franchisees below:

Franchise	Franchisee	Trading Name
Chiltern	Deutsche Bahn (Arriva) (until 31 December 2021)	**Chiltern Railways**

Chiltern Railways operates a frequent service between London Marylebone, Banbury and Birmingham Snow Hill, with some peak trains extending to Kidderminster. There are also regular services from Marylebone to Stratford-upon-Avon and to Aylesbury Vale Parkway via Amersham (along the London Underground Metropolitan Line). The fleet consists of DMUs of Classes 121 (used on the Princes Risborough–Aylesbury route), 165, 168 and 172 plus a number of loco-hauled rakes used on some of the Birmingham route trains, worked by Class 67s hired from DB Schenker.

| **Cross-Country** | Deutsche Bahn (Arriva) | **CrossCountry** |
| | (until 31 March 2016)* | |

Franchise extension to be negotiated to November 2019.

CrossCountry operates a network of long distance services between Scotland, North-East England and Manchester to the South-West of England, Reading, Southampton, Bournemouth and Guildford, centred on Birmingham New Street. These trains are mainly formed of diesel Class 220/221 Voyagers, supplemented by a small number of HSTs on the NE–SW route. Inter-urban services also link Nottingham, Leicester and Stansted Airport with Birmingham and Cardiff. These use Class 170 DMUs.

| **East Midlands** | Stagecoach Group plc | **East Midlands Trains** |
| | (until 31 March 2015)* | |

Franchise extension to be negotiated to October 2017.

EMT operates a mix of long distance high speed services on the Midland Main Line (MML), from London St Pancras to Sheffield (Leeds at peak times and some extensions to York/Scarborough) and Nottingham, and local and regional services ranging from the Norwich–Liverpool route to Nottingham–Skegness, Nottingham–Mansfield–Worksop, Nottingham–Matlock and Derby–Crewe. It also operates local services in Lincolnshire. Trains on the MML are worked by a fleet of Class 222 DMUs and ten HSTs, whilst the local and regional fleet consists of DMU Classes 153, 156 and 158.

| **Essex Thameside** | National Express Group plc | **c2c** |
| | (until 8 November 2029) | |

c2c operates an intensive, principally commuter, service from London Fenchurch Street to Southend and Shoeburyness via both Upminster and Tilbury. The fleet consists entirely of Class 357 EMUs. In 2014 c2c won a new 15-year franchise that promised to introduce 17 new 4-car EMUs from 2019.

| **Greater Western** | First Group plc | **First Great Western** |
| | (until 20 September 2015) | |

Franchise extension currently being negotiated.

First Great Western operates long distance trains from London Paddington to South Wales, the West Country and Worcester and Hereford. In addition there are frequent trains along the Thames Valley corridor to Newbury and Oxford, plus local and regional trains throughout the South-West including the Cornish, Devon and Thames Valley branches, the Reading–Gatwick North Downs line and Cardiff–Portsmouth Harbour and Bristol–Weymouth regional routes. A fleet of 53 HSTs is used on the long-distance trains, with DMUs of Classes 165 and 166 used on the North Downs and Thames Valley routes and Class 180s used alongside HSTs on the Cotswold Line to Worcester and Hereford. Classes 143, 150, 153 and 158 are used on local and regional trains in the South-West. A small fleet of four Class 57s is maintained to work the overnight "Cornish Riviera" Sleeper service between London and Penzance.

| **Greater Anglia** | Abellio (NS) | **Abellio Greater Anglia** |
| | (until 19 October 2016) | |

Abellio Greater Anglia operates main line trains between London Liverpool Street, Ipswich and Norwich and local trains across Norfolk, Suffolk and parts of Cambridgeshire. It also runs local and commuter services into Liverpool Street from the Great Eastern (including Southend, Braintree and Clacton) and West Anglia (including Cambridge and Stansted Airport) routes. It operates a varied fleet of Class 90s with loco-hauled Mark 3 sets, DMUs of Classes 153, 156 and 170 and EMUs of Classes 315, 317, 321, 360 and 379. London Overground is due to take over some suburban services from Liverpool Street (to Chingford, Cheshunt and Enfield) in May 2015.

| **Integrated Kent** | Govia Ltd (Go-Ahead/Keolis) (until 24 June 2018) | **Southeastern** |

Southeastern operates all services in the South-East London suburbs, the whole of Kent and part of Sussex, which are primarily commuter services to London. It also operates domestic high speed trains on HS1 from St Pancras to Ashford, Ramsgate, Dover and Faversham with additional peak services on other routes. EMUs of Classes 375, 376, 465 and 466 are used, along with Class 395s on the High Speed trains.

| **InterCity East Coast** | Directly Operated Railways (until 31 March 2015) | **East Coast** |

Currently run on an interim basis by DfT management company Directly Operated Railways (trading as East Coast). This arrangement is due to continue until a new franchise is let to the private sector, with the new franchise currently planned to start on 1 April 2015.

East Coast operates frequent long distance trains on the East Coast Main Line between London King's Cross, Leeds, York, Newcastle and Edinburgh, with less frequent services to Bradford, Harrogate, Skipton, Hull, Lincoln, Glasgow, Aberdeen and Inverness. A mixed fleet of Class 91s and 30 Mark 4 sets, and 14 HST sets, are used on these trains.

| **InterCity West Coast** | Virgin Rail Group Ltd (until 31 March 2017) | **Virgin Trains** |

Virgin operates long distance services along the West Coast Main Line from London Euston to Birmingham/Wolverhampton, Manchester, Liverpool and Glasgow using Class 390 Pendolino EMUs. It also operates Class 221 Voyagers on the Euston–Chester–Holyhead route, whilst a mixture of 221s and 390s are used on the Euston–Birmingham–Glasgow/Edinburgh route. One rake of Mark 3 loco-hauled stock is also leased, and normally used on Thursdays and Fridays (this was due to be returned to Porterbrook at the end of October 2014).

| **London Rail** | MTR/Deutsche Bahn (until 12 November 2016) | **London Overground** |

This is a Concession and is different from other rail franchises, as fares and service levels are set by Transport for London instead of by the DfT.

London Overground operates services on the Richmond–Stratford North London Line and the Willesden Junction–Clapham Junction West London Line, plus the new East London Line from Highbury & Islington to New Cross and New Cross Gate, with extensions to Clapham Jn (via Denmark Hill), Crystal Palace and West Croydon. It also runs services from London Euston to Watford Junction. All these use Class 378 EMUs whilst Class 172 DMUs are used on the Gospel Oak–Barking route.

| **Merseyrail Electrics** | Serco/Abellio (NS) (until 19 July 2028) | **Merseyrail** |

Under the control of Merseytravel PTE instead of the DfT. Franchise reviewed every five years to fit in with the Merseyside Local Transport Plan.

Merseyrail operates services between Liverpool and Southport, Ormskirk, Kirkby, Hunts Cross, New Brighton, West Kirby, Chester and Ellesmere Port, all worked by EMUs of Classes 507 and 508.

| **Northern Rail** | Serco/Abellio (NS) | **Northern** |
| | (until 1 February 2016) | |

Northern operates a range of inter-urban, commuter and rural services throughout the North of England, including those around the cities of Leeds, Manchester, Sheffield, Liverpool and Newcastle. The network extends from Chathill in the north to Nottingham in the south, and Cleethorpes in the east to St Bees in the west. Long distance services include Leeds–Carlisle, Middlesbrough–Carlisle and York–Blackpool North. The operator uses a large fleet of DMUs of Classes 142, 144, 150, 153, 155, 156 and 158 plus EMU Classes 321, 322, 323 and 333.

| **ScotRail** | First Group plc | **ScotRail** |
| | (until 31 March 2015) | |

Abellio has won the contract to operate the ScotRail franchise from April 2015.

ScotRail provides almost all passenger services within Scotland and also trains from Glasgow to Carlisle via Dumfries, some of which extend to Newcastle (jointly operated with Northern). The company also operates the overnight Caledonian Sleeper services between London and Glasgow, Edinburgh, Inverness, Aberdeen and Fort William. In addition to the Sleeper loco-hauled stock (hauled by Class 67s and 90s hired from DB Schenker), the company operates a large fleet of DMUs of Classes 156, 158 and 170 and EMU Classes 314, 318, 320, 334 and 380. One loco-hauled rake is also used on a Fife Circle commuter train, hauled by a Class 67.

| **South Central** | Govia Ltd (Go-Ahead/Keolis) | **Southern** |
| | (until 25 July 2015) | |

Upon termination of the Southern franchise in July 2015 it is to be combined with the new Thameslink, Southern & Great Northern franchise (also operated by Govia).

Southern operates predominantly commuter services between London, Surrey and Sussex, as well as services along the South Coast between Southampton, Brighton, Hastings and Ashford, and also the cross-London service linking South Croydon and Milton Keynes. It also operates metro services in South London and Gatwick Express, which is a premium non-stop service between London Victoria and Gatwick Airport. Class 171 DMUs are used on Brighton–Ashford and London Bridge–Uckfield services, whilst all other services are in the hands of EMUs of Classes 313, 377, 442 and 455.

| **South Western** | Stagecoach Group plc | **South West Trains** |
| | (until 3 February 2017)* | |

Franchise extension to be negotiated to April 2019.

South West Trains operates trains from London Waterloo to destinations across the South and South-West including Woking, Basingstoke, Southampton, Portsmouth, Salisbury, Exeter, Reading and Weymouth as well as suburban services from Waterloo. SWT also runs services between Ryde and Shanklin on the Isle of Wight, using former London Underground 1938 stock (Class 483). The rest of the fleet consists of DMU Classes 158 and 159 and EMU Classes 444, 450, 455, 456 and 458.

| **Thameslink & Great Northern** | Govia Ltd (Go-Ahead/Keolis) | **Govia Thameslink Railway** |
| | (until 19 September 2021) | |

The Southern franchise will be combined with Govia Thameslink Railway from July 2015.

Govia operates this franchise as a management contract. GTR operates trains on the Thameslink route between Bedford and Brighton via central London and also on the Sutton and Wimbledon loops. A joint service with Southeastern is also operated to Sevenoaks, Orpington and Ashford. GTR also runs services on the Great Northern route from London King's Cross and Moorgate to Welwyn Garden City, Hertford North, Peterborough, Cambridge and Kings Lynn. The fleet consists of EMU Classes 319 and 377 (and 387 from 2015) for the Thameslink route services and Classes 313, 317, 321 and 365 for the Great Northern route services.

Trans-Pennine Express   First Group/Keolis   **TransPennine Express**
(until 1 April 2015).
Franchise extension to be negotiated to February 2016.

TransPennine Express operates predominantly long distance inter-urban services linking major cities across the North of England, along with Edinburgh and Glasgow in Scotland. The main services are Manchester Airport/Manchester Piccadilly–Newcastle/Middlesbrough/Hull plus Liverpool–Scarborough and Liverpool–Newcastle along the North Trans-Pennine route via Huddersfield, Leeds and York, and Manchester Airport–Cleethorpes along the South Trans-Pennine route via Sheffield. TPE also operates Manchester Airport–Blackpool/Barrow/Windermere/Edinburgh/Glasgow. The fleet consists of DMU Classes 170 (used on the Hull and Cleethorpes routes) and 185 and also new Class 350 EMUs used on Manchester Airport–Scotland services.

Wales & Borders   Deutsche Bahn (Arriva)   **Arriva Trains Wales**
(until 14 October 2018)*
The franchise agreement includes the provision for the term to be further extended by mutual agreement by up to five years beyond October 2018. Management of the franchise is devolved to the Welsh Government, but DfT is still the procuring authority.

Arriva Trains Wales operates a mix of long distance, regional and local services throughout Wales, including the Valley Lines network of lines around Cardiff, and also through services to the English border counties and to Manchester and Birmingham. The fleet consists of DMUs of Classes 142, 143, 150, 158 and 175 and one loco-hauled rake used on a premium Welsh Government sponsored service on the Cardiff–Holyhead route, hauled by a Class 67.

West Midlands   Govia Ltd (Go-Ahead/Keolis)   **London Midland**
(until 19 September 2015)*
Franchise extension to be negotiated to June 2017.

London Midland operates long distance and regional services from London Euston to Northampton and Birmingham/Crewe and also between Birmingham and Liverpool as well as local and regional services around Birmingham, including to Stratford-upon-Avon, Worcester, Hereford, Redditch and Shrewsbury. It also operates the Bedford–Bletchley and Watford Jn–St Albans Abbey branches. The fleet consists of DMU Classes 150, 153, 170 and 172 and EMU Classes 321, 323 and 350.

* Franchise agreement includes provision for an extension of up to seven 4-week reporting periods.

The following operators run non-franchised services (* special summer services only):

Operator	Trading Name	Route
BAA	Heathrow Express	London Paddington–Heathrow Airport
First Hull Trains	First Hull Trains	London King's Cross–Hull
Grand Central	Grand Central	London King's Cross–Sunderland/Bradford Interchange
North Yorkshire Moors Railway Enterprises	North Yorkshire Moors Railway	Pickering–Grosmont–Whitby/Battersby
West Coast Railway Company	West Coast Railway Company	Birmingham–Stratford-upon-Avon* Fort William–Mallaig* York–Wakefield–York–Scarborough*

INTERNATIONAL PASSENGER OPERATIONS

Eurostar operates passenger services between the UK and mainland Europe.

Eurostar International, established in 2010, is jointly owned by the SNCF (the national operator of France), 55%, SNCB (the national operator of Belgium), 5% and HM Treasury, 40%. The 40% UK stake was transferred from London & Continental Railways (LCR) to HM Treasury in 2014. LCR had bought British Rail's interest in Eurostar at the time of the UK railway privatisation in 1996.

In addition, a service for the conveyance of accompanied road vehicles through the Channel Tunnel is provided by the tunnel operating company, Eurotunnel.

FREIGHT TRAIN OPERATIONS

The following operators operate freight services or empty passenger stock workings under "Open Access" arrangements:

Colas Rail: In addition to its On-Track Machines and infrastructure activities, Colas Rail operates a number of freight flows, including steel, coal and timber. Colas Rail has a small but varied fleet consisting of Class 37s, 47s, 56s, 60s, 66s and 70s. The ten Class 60s were acquired from DBS in 2014 and initially five are being returned to service.

DB Schenker Rail (UK): Still the biggest freight operator in the country, DBS (formerly EWS before being bought by DB) has seen some of its core traffic lost to competitors in recent years. It still provides a large number of infrastructure trains to Network Rail and operates coal, steel, intermodal and aggregate trains nationwide. The core fleet is Class 66s. Of the original 250 ordered 176 are still used in the UK, with the remainder having moved to DB's French and Polish operations, although some of the French locos do return to the UK when major maintenance is required. A fleet of around 25 Class 60s are also used on heavier trains.

DBS's six Class 59/2s are used alongside the Mendip Rail 59/0s and 59/1s on stone traffic from the Mendip quarries and around the South-East. DBS's fleet of Class 67s are mainly used on passenger or standby duties for Arriva Trains Wales, Chiltern Railways, East Coast and ScotRail. Class 90s are hired to ScotRail for use on the sleeping car services but see little use on freight, whilst the fleet of Class 92s are mainly used on intermodal duties, including a limited number of overnight trains on High Speed 1.

DBS also operates a number of excursion trains.

Devon & Cornwall Railways (a subsidiary of British American Railway Services): DCRail specialises in short-term freight haulage contracts, mainly in the scrap, coal and aggregates markets, using its fleet of Class 56s. It also provides locomotives from its fleet of 31s or 56s for stock moves and has some contracts with Network Rail for moving On Track Machines or other equipment.

Direct Rail Services: DRS has built on its original nuclear flask traffic to operate a number of different services. The main flows are intermodal plus the provision of crews and locos to Network Rail for autumn Railhead

Treatment Trains. Its Class 47s and 57s are also used on excursion work. DRS has the most varied fleet of locomotives, with Class 20s, 37s, 47s, 57s and 66s working alongside a fleet of Class 68s that are currently being delivered. The company has 25 Class 68s on order as well as ten new Vossloh electric locos (Class 88s), that will also feature a small diesel engine.

DRS also operates a number of excursion trains.

Freightliner: Freightliner has two divisions: Intermodal operates container trains from the main Ports at Southampton, Felixstowe, Tilbury and Thamesport to major cities including London, Manchester, Leeds and Birmingham. The Heavy Haul division covers the movement of coal, cement, infrastructure and aggregates nationwide. Most services are worked by Class 66s, with Class 70s used on some of the heavier intermodal trains and some Heavy Haul flows, principally coal, cement and ballast trains. A small fleet of Class 86 and 90 electrics is used on intermodal trains on the Great Eastern and West Coast Main Lines, the Class 86s mainly being used in pairs on the WCML between Crewe and Coatbridge.

GB Railfreight: GBRf, owned by Eurotunnel, operates a mixture of traffic types, mainly using Class 66s together with a small fleet of Class 73s on infrastructure duties in the South-East. A growing fleet of Class 92s is also used on some intermodal flows to/from Dollands Moor. Traffic includes coal, intermodal, biomass and gypsum as well as infrastructure services for Network Rail and London Underground.

GBRf also operates some excursion trains, including those using the preserved Class 201 "Hastings" DEMU.

West Coast Railway Company: WCRC has a freight licence but doesn't operate any freight as such – only empty stock movements. Its fleet of 47s, supplemented by a smaller number of 33s, 37s and 57s, is used on excursion work nationwide, including the prestigious Royal Scotsman.

In addition a number of other companies operate infrastructure trains, mainly formed of On-Track Machines. These include Balfour Beatty Rail Infrastructure Services, Swietelsky Babcock Rail, Volker Rail and South West Trains.

INTRODUCTION

COACHING STOCK CLASSIFICATION

Seven different numbering systems were in use on British Rail. These were the British Rail series, the four pre-nationalisation companies' series', the Pullman Car Company's series and the UIC (International Union of Railways) series. In this book BR number series coaches and former Pullman Car Company series are listed separately. There is also a separate listing of "Saloon" type vehicles, that includes pre-nationalisation survivors, which are registered to run on the national railway system, Locomotive Support Coaches and Service Stock. Please note the Mark 2 Pullman vehicles were ordered after the Pullman Car Company had been nationalised and are therefore numbered in the British Rail series.

Also listed separately are the British Rail and Pullman Car Company number series coaches used on North Yorkshire Moors Railway services on the national railway system. This is due to their very restricted sphere of operation.

The BR number series grouped vehicles of a particular type together in chronological order. Major modifications affecting type of accommodation resulted in renumbering into a more appropriate or new number series. Since privatisation such renumbering has not always taken place resulting in renumbering which has been more haphazard and greater variations within numbering groups.

LAYOUT OF INFORMATION

Coaches are listed in numerical order of painted number in batches according to type.

Each coach entry is laid out as in the following example. Where a coach has been renumbered, the former number is shown in parentheses. If a coach has been renumbered more than once, the original number is shown first in parentheses, followed by the most recent previous number.

No.	Prev. No.	Notes	Livery	Owner	Operator	Depot/Location
42346	(41053)	*h	**FD**	A	*GW*	LA

The owner is the responsible custodian of the coach and this may not always be the legal owner.

The operator is the organisation which facilitates the use of the coach and may not be the actual train operating company which runs the train. If no operator is shown the coach is considered to be not in use.

The depot is the facility primarily responsible for the coaches maintenance. Light maintenance and heavy overhauls may be carried out elsewhere.

The location is where coaches not in use are currently being kept or stored.

Detailed information & codes: Under each type heading, the following details are shown:

- "Mark" of coach (see below).
- Descriptive text.
- Number of First Class seats, Standard Class seats, lavatory compartments and wheelchair spaces shown as F/S nT nW respectively.
- Bogie type (see below).
- Additional features.
- ETH Index.
- Weight

GENERAL INFORMATION

The number of locomotive-hauled or propelled carriages in use on the national railway network is much fewer than was once the case and their number is expected to reduce further as more new multiple units are delivered. Those that remain fall into two distinct groups.

Firstly, there are those used by franchised and open access operators for regular timetabled services. Most of these are formed in fixed or semi-fixed formations with either locomotives or a locomotive and Driving Brake Van at either end which allows for push-pull operation. There are also a small number of mainly overnight trains with variable formations that use conventional locomotive haulage.

Secondly there are those used for what can best be described as excursion trains. These include a wide range of carriage types ranging from luxurious saloons to those more suited to the "bucket and spade" seaside type of excursion. These are formed into sets to suit the requirements of the day. From time to time some see limited use with franchised and open access operators to cover for stock shortages and times of exceptional demand such as major sporting events.

In addition there remain a small number of carriages referred to as "Service Stock" which are used internally within the railway industry and are not used to convey passengers.

FRANCHISED & OPEN ACCESS OPERATORS

For each operator regularly using locomotive-hauled carriages brief details are given here of the sphere if operation. For details of operators using HSTs see Section 2.1.

Arriva Trains Wales
The daily Welsh Assembly Government sponsored train between Cardiff and Holyhead uses Mark 3 carriages in push-pull mode with a Class 67. These carriages are also used for relief trains, particularly in connection with sports fixtures at Cardiff and busy ferry sailings to/from Holyhead. Additional carriages are currently being overhauled for regular use on North Wales services.

Chiltern Railways
Chiltern operates four sets of Mark 3 carriages with Class 67 locomotives (to be replaced with Class 68s in 2015) on its Mainline services between London Marylebone and Birmingham Moor Street/Kidderminster. Another set is used on a peak-hour commuter service between Marylebone and Banbury. All trains operate as push-pull sets.

East Coast
East Coast operates 30 sets of Mark 4 carriages with Class 91 locomotives in push-pull formations on its Inter-City services between London King's Cross and Yorkshire, North-East England and Scotland.

First Great Western
The "Night Riviera" seating and sleeping car service between London Paddington and Penzance uses sets of Mark 3 carriages hauled by Class 57/6 locomotives. The seating carriages are also used in Devon and Cornwall for local services on summer Saturdays.

Greater Anglia
The Inter-City service between London Liverpool Street and Norwich is operated using 12 sets of Mark 3 carriages with Class 90 locomotives in push-pull formations. On summer Saturdays some of these trains are extended to Great Yarmouth hauled beyond Norwich by Class 47 s. In addition local trains between Norwich and Great Yarmouth or Lowestoft are occasionally Class 47 hauled using spare Mark 3s or hired excursion carriages.

North Yorkshire Moors Railway
In addition to operating the North Yorkshire Moors Railway between Pickering and Grosmont the company operates through services to Whitby and occasionally Battersby. A fleet of Mark 1 passenger carriages and Pullman Cars are used for these services.

ScotRail
The "Caledonian Sleeper" seating and sleeping car service between London Euston and Scotland uses sets of Mark 3 Sleeping Cars and Mark 2 seating and catering carriages. These are hauled by Class 90 locomotives between London and Edinburgh/Glasgow and Class 67 locomotives between Edinburgh and Aberdeen, Inverness and Fort William. In addition, a Class 67 is used to haul a hired set of Mark 2 excursion stock on peak hour services between Edinburgh and Fife.

Virgin Trains
Virgin retains a set of Mark 3 carriages operated with a Class 90 locomotive in push-pull formation that is used for relief trains on the West Coast Main Line. The set also sees occasional excursion use. This set is due to be returned to Porterbrook at the end of October 2014 and will not be replaced.

EXCURSION TRAIN OPERATORS

Usually, three types of companies will be involved in the operation of an excursion train. There will be the promoter, the rolling stock provider and the train operator. In many cases two or more of these roles may be undertaken by the same or associated companies. Only a small number of Train Operating Companies facilitate the operation of excursion trains. This takes various forms ranging from the complete package of providing and operating the train, through offering a "hook up and haul" service, to operating the train for a third party rolling stock custodian.

DB Schenker Rail UK
DBS currently operates its own luxurious train of Mark 3 carriages, called its company train. Otherwise it offers a hook up and haul service and regularly operates the Royal Train and the Belmond British Pullman as well as trains for Riviera Trains and its client promoters. An excursion train fleet of Mark 2 carriages is owned but only a very small number of these currently see use being hired to ScotRail for Edinburgh–Fife peak-hour trains.

Direct Rail Services
DRS operates a small fleet of Mark 2 carriages which see occasional use on excursion trains. These are also hired to franchised operators to cover stock shortages or periods of exceptional passenger demand. The company also offers a hook up and haul service operating the Belmond Northern Belle as well as trains for Riviera Trains and their client promoters.

West Coast Railway Company
This vertically integrated company has its own fleet of steam and diesel locomotives as well as a full range of different carriage types. It operates its own regular trains, such as the "Jacobite" steam service between Fort William and Mallaig and numerous excursion trains for itself and client promoters. In addition it offers a hook up and haul service operating the Royal Scotsman and Statesman trains as well as trains for Vintage Trains, The Princess Royal Locomotive Trust and the Scottish Railway Preservation Society.

BOGIE TYPES

BR Mark 1 (BR1). Double bolster leaf spring bogie. Generally 90 mph, but Mark 1 bogies may be permitted to run at 100 mph with special maintenance. Weight: 6.1 t.

BR Mark 2 (BR2). Single bolster leaf-spring bogie used on certain types of non-passenger stock and suburban stock (all now withdrawn). Weight: 5.3 t.

COMMONWEALTH (C). Heavy, cast steel coil spring bogie. 100 mph. Weight: 6.75 t.

B4. Coil spring fabricated bogie. Generally 100 mph, but B4 bogies may be permitted to run at 110 mph with special maintenance. Weight: 5.2 t.

B5. Heavy duty version of B4. 100 mph. Weight: 5.3 t.

B5 (SR). A bogie originally used on Southern Region EMUs, similar in design to B5. Now also used on locomotive-hauled coaches. 100 mph.

BT10. A fabricated bogie designed for 125 mph. Air suspension.

T4. A 125 mph bogie designed by BREL (now Bombardier Transportation).

BT41. Fitted to Mark 4 vehicles, designed by SIG in Switzerland. At present limited to 125 mph, but designed for 140 mph.

BRAKES

Air braking is now standard on British main line trains. Vehicles with other equipment are denoted:

b   Air braked, through vacuum pipe.
v   Vacuum braked.
x   Dual braked (air and vacuum).

HEATING & VENTILATION

Electric heating and ventilation is now standard on British main-line trains. Certain coaches for use on excursion services may also have steam heating facilities, or be steam heated only. All carriages used on North Yorkshire Moors Railway trains have steam heating.

PUBLIC ADDRESS

It is assumed all coaches are now fitted with public address equipment, although certain stored vehicles may not have this feature. In addition, it is assumed all vehicles with a conductor's compartment have public address transmission facilities, as have catering vehicles.

COOKING EQUIPMENT

It is assumed that Mark 1 catering vehicles have gas powered cooking equipment, whilst Mark 2, 3 and 4 catering vehicles have electric powered cooking equipment unless stated otherwise.

ADDITIONAL FEATURE CODES

d	Central Door Locking.
dg	Driver–Guard communication equipment.
f	Facelifted or fluorescent lighting.
h	"High density" seating
k	Composition brake blocks (instead of cast iron).
n	Day/night lighting.
pg	Public address transmission and driver-guard communication.
pt	Public address transmission facility.
q	Catering staff to shore telephone.
w	Wheelchair space.
★	Blue star multiple working cables fitted.

NOTES ON ETH INDICES

The sum of ETH indices in a train must not be more than the ETH index of the locomotive. The normal voltage on British trains is 1000 V. Suffix "X" denotes 600 amp wiring instead of 400 amp. Trains whose ETH index is higher than 66 must be formed completely of 600 amp wired stock. Class 33 and 73 locomotives cannot provide a suitable electric train supply for Mark 2D, Mark 2E, Mark 2F, Mark 3, Mark 3A, Mark 3B or Mark 4 coaches. Class 55 locomotives provide an ets directly from one of their traction generators into the train line. Consequently voltage fluctuations can result in motor-alternator flashover. Thus these locomotives are not suitable for use with Mark 2D, Mark 2E, Mark 2F, Mark 3, Mark 3A, Mark 3B or Mark 4 coaches unless modified motor-alternators are fitted. Such motor alternators were fitted to Mark 2D and 2F coaches used on the East Coast Main Line, but few remain fitted.

BUILD DETAILS

Lot Numbers
Vehicles ordered under the auspices of BR were allocated a lot (batch) number when ordered and these are quoted in class headings and sub-headings.

Builders
These are shown in class headings, the following designations being used:

Ashford	BR, Ashford Works.
BRCW	Birmingham Railway Carriage & Wagon Company, Smethwick, Birmingham.
BREL Derby	BREL, Derby Carriage Works (later ABB/Adtranz Derby, now Bombardier Transportation Derby).
Charles Roberts	Charles Roberts & Company, Horbury, Wakefield (later Bombardier Transportation).
Cravens	Cravens, Sheffield.
Derby	BR, Derby Carriage Works (later BREL Derby, then ABB/Adtranz Derby, now Bombardier Transportation Derby).

Doncaster	BR, Doncaster Works (later BREL Doncaster, then BRML Doncaster, then ABB/Adtranz Doncaster, then Bombardier).
Eastleigh	BR, Eastleigh Works (later BREL Eastleigh, then Wessex Traincare, and Alstom Eastleigh, now Arlington Fleet Services).
Glasgow	BR Springburn Works, Glasgow (now Knorr-Bremse Rail Systems).
Gloucester	The Gloucester Railway Carriage & Wagon Co.
Hunslet-Barclay	Hunslet Barclay, Kilmarnock Works (now Wabtec Rail Scotland).
Metro-Cammell	Metropolitan-Cammell, Saltley, Birmingham (later GEC-Alsthom Birmingham, then Alstom Birmingham).
Pressed Steel	Pressed Steel, Linwood.
Swindon	BR Swindon Works.
Wolverton	BR Wolverton Works (later BREL Wolverton then Railcare, Wolverton, then Alstom, Wolverton now Knorr-Bremse Rail Systems).
York	BR, York Carriage Works (later BREL York, then ABB York).

Information on sub-contracting works which built parts of vehicles eg the underframes etc is not shown. In addition to the above, certain vintage Pullman cars were built or rebuilt at the following works:

Metropolitan Carriage & Wagon Company, Birmingham (later Alstom).
Midland Carriage & Wagon Company, Birmingham.
Pullman Car Company, Preston Park, Brighton.
Conversions have also been carried out at the Railway Technical Centre, Derby, LNWR, Crewe and Blakes Fabrications, Edinburgh.

ABBREVIATIONS

The following abbreviations are used in class headings and also throughout this publication:

BR	British Railways.
DB	Deutsche Bahn
DEMU	Diesel Electric Multiple Unit.
DMU	Diesel Multiple Unit (general term).
EMU	Electric Multiple Unit.
ETH	Electric Train Heating
ft	feet
GWR	Great Western Railway
kN	kilonewtons.
km/h	kilometres per hour.
kW	kilowatts.
LT	London Transport.
LUL	London Underground Limited.
m	metres.
mph	miles per hour.
SR	BR Southern Region
t	tonnes.

▲ Pullman Car Company-liveried Mark 2 Pullman Open First 550 "RYDAL WATER" at Kirkby-in-Furness on 13/04/13. **Andrew Mason**

▼ Riviera Trains Oxford blue-liveried Mark 1 Kitchen Buffet Unclassified 1683 near Scarborough on 10/05/14. **Andrew Mason**

▲ BR maroon-liveried Mark 1 Buffet Standard 1882 (carrying the number 99311) at Stoneycombe on 09/06/13. **Tony Christie**

▼ HRH The Prince of Wales's Sleeping Car Mark 3B 2922 at Wolverton on 23/03/12. **Mark Beal**

▲ BR Carmine & Cream-liveried Mark 1 Open First 3097 at Soulbury on 07/08/14.
Mark Beal

▼ BR Western Region/GWR chocolate & cream-liveried Mark 1 Open First 3125 at Exeter St Davids on 25/08/14.
Tony Christie

▲ Riviera Trains Great Briton-liveried Mark 2F Open First 3348 "GAINSBOROUGH" at Docker on 17/05/14. **Andrew Mason**

▼ BR maroon-liveried Mark 1 Open Standard 3766 (carrying the number 99317) at Worcester Shrub Hill on 07/06/14. **Steve Widdowson**

▲ BR Western Region/GWR chocolate & cream-liveried Mark 2 Open Standard 5212 near Scarborough on 12/07/14. **Andrew Mason**

▼ BR maroon-liveried Mark 2D Open Standard 5657 near Inverkeithing on 09/07/14. **Ian Lothian**

▲ First Group-liveried Mark 2E Open Brake Unclassified 9804, used in the ScotRail Caledonian Sleeper trains, at Carluke on 20/06/14. **Robin Ralston**

▼ Chiltern Mainline-liveried Mark 3A Kitchen Buffet First 10274 at Birmingham Moor Street on 05/09/14. **Robert Pritchard**

▲ First Group Dynamic Lines-liveried Mark 3A Sleeping Car With Pantry 10532 at Plymouth on 06/08/14. **Tony Christie**

▼ Abellio Greater Anglia-liveried Mark 3B Open First 11096 at Stratford on 29/04/14. **Robert Pritchard**

▲ East Coast-liveried Mark 4 Open Standard 12465 at Doncaster on 09/09/14.
Robert Pritchard

▼ Revised Direct Rail Services-liveried Mark 2D Corridor Brake First 17159 at North Staffs Junction on 26/08/13.
Andrew Mason

THE DEVELOPMENT OF BR STANDARD COACHES

Mark 1

The standard BR coach built from 1951 to 1963 was the Mark 1. This type features a separate underframe and body. The underframe is normally 64 ft 6 in long, but certain vehicles were built on shorter (57 ft) frames. Tungsten lighting was standard and until 1961, BR Mark 1 bogies were generally provided. In 1959 Lot No. 30525 (Open Standard) appeared with fluorescent lighting and melamine interior panels, and from 1961 onwards Commonwealth bogies were fitted in an attempt to improve the quality of ride which became very poor when the tyre profiles on the wheels of the BR1 bogies became worn. Later batches of Open Standard and Open Brake Standard retained the features of Lot No. 30525, but compartment vehicles – whilst utilising melamine panelling in Standard Class – still retained tungsten lighting. Wooden interior finish was retained in First Class vehicles where the only change was to fluorescent lighting in open vehicles (except Lot No. 30648, which had tungsten lighting). In later years many Mark 1 coaches had BR 1 bogies replaced by B4.

XP64

In 1964, a new prototype train was introduced. Known as "XP64", it featured new seat designs, pressure heating & ventilation, aluminium compartment doors and corridor partitions, foot pedal operated toilets and B4 bogies. The vehicles were built on standard Mark 1 underframes. Folding exterior doors were fitted, but these proved troublesome and were later replaced with hinged doors. All XP64 coaches have been withdrawn, but some have been preserved.

Mark 2

The prototype Mark 2 vehicle (W13252) was produced in 1963. This was a Corridor First of semi-integral construction and had pressure heating & ventilation, tungsten lighting, and was mounted on B4 bogies. This vehicle has now been preserved at the Mid Norfolk Railway. The production build was similar, but wider windows were used. The Open Standard vehicles used a new seat design similar to that in the XP64 and fluorescent lighting was provided. Interior finish reverted to wood. Mark 2 vehicles were built from 1964–66.

Mark 2A–2C

The Mark 2A design, built 1967–68, incorporated the remainder of the features first used in the XP64 coaches, ie foot pedal operated toilets (except Open Brake Standard), new First Class seat design, aluminium compartment doors and partitions together with fluorescent lighting in first class compartments. Folding gangway doors (lime green coloured) were used instead of the traditional one-piece variety.

Mark 2B coaches had wide wrap around doors at vehicle ends, no centre doors and a slightly longer body. In Standard Class there was one toilet at each end instead of two at one end as previously. The folding gangway doors were red.

Mark 2C coaches had a lowered ceiling with twin strips of fluorescent lighting and ducting for air conditioning, but air conditioning was never fitted.

Mark 2D–2F

These vehicles were fitted with air conditioning. They had no opening top-lights in saloon windows, which were shallower than previous ones.

Mark 2E vehicles had smaller toilets with luggage racks opposite. The folding gangway doors were fawn coloured.

Mark 2F vehicles had a modified air conditioning system, plastic interior panels and InterCity 70 type seats.

Mark 3

The Mark 3 design has BT10 bogies, is 75 ft (23 m) long and is of fully integral construction with InterCity 70 type seats. Gangway doors were yellow (red in Kitchen Buffet First) when new, although these were changed on refurbishment. Loco-hauled coaches are classified Mark 3A, Mark 3 being reserved for HST trailers. A new batch of Open First and Open Brake First, classified Mark 3B, was built in 1985 with Advanced Passenger Train-style seating and revised lighting. The last vehicles in the Mark 3 series were the driving brake vans ("Driving Van Trailers") built for West Coast Main Line services but now used elsewhere.

A number of Mark 3 vehicles have recently been converted for use as HST trailers with Grand Central, CrossCountry and First Great Western.

Mark 4

The Mark 4 design was built by Metro-Cammell for use on the East Coast Main Line after electrification and featured a body profile suitable for tilting trains, although tilt is not fitted, and is not intended to be. This design is suitable for 140 mph running, although is restricted to 125 mph because the signalling system on the route is not suitable for the higher speed. The bogies for these coaches were built by SIG in Switzerland and are designated BT41. Power operated sliding plug exterior doors are standard. All Mark 4s were rebuilt with completely new interiors in 2003–05 for GNER and referred to as "Mallard" stock. These rakes generally run in fixed formations and are now operated by East Coast.

1. BR NUMBER SERIES COACHING STOCK

KITCHEN FIRST

Mark 1. Spent most of its life as a Royal Train vehicle and was numbered 2907 for a time. 24/–. Built with Commonwealth bogies, but B5 bogies substituted. ETH 2.

Lot No. 30633 Swindon 1961. 41 t.

325	**VN**	BE	*NB*	CP	DUART

PULLMAN KITCHEN

Mark 2. Pressure Ventilated. Built with First Class seating but this has been replaced with a servery area. 2T. B5 bogies. ETH 6.

Lot No. 30755 Derby 1966. 40 t.

504	**PC**	WC	*WC*	CS	ULLSWATER
506	**PC**	WC	*WC*	CS	WINDERMERE

PULLMAN OPEN FIRST

Mark 2. Pressure Ventilated. 36/– 2T. B4 bogies. ETH 5.

Lot No. 30754 Derby 1966. 35 t.

Non-standard livery: 546 Maroon & beige.

546	**0**	WC		CS	CITY OF MANCHESTER
548	**PC**	WC	*WC*	CS	GRASMERE
549	**PC**	WC	*WC*	CS	BASSENTHWAITE
550	**PC**	WC	*WC*	CS	RYDAL WATER
551	**PC**	WC	*WC*	CS	BUTTERMERE
552	**PC**	WC	*WC*	CS	ENNERDALE WATER
553	**PC**	WC	*WC*	CS	CRUMMOCK WATER

PULLMAN OPEN BRAKE FIRST

Mark 2. Pressure Ventilated. 30/– 2T. B4 bogies. ETH 4.

Lot No. 30753 Derby 1966. 35 t.

586	**PC**	WC	*WC*	CS	DERWENTWATER

BUFFET FIRST

Mark 2F. Air conditioned. Converted 1988–89/91 at BREL, Derby from Mark 2F Open Firsts. 1200/01/03/11/20/21 have Stones equipment, others have Temperature Ltd. 25/– 1T 1W. B4 bogies. d. ETH 6X.

1200/03/11/20. Lot No. 30845 Derby 1973. 33 t.
1201/07/10/12/21/54. Lot No. 30859 Derby 1973–74. 33 t.

1200	(3287, 6459)	**RV** RV	*RV*	EH	AMBER
1201	(3361, 6445)	**CH** VT	*WC*	CS	
1203	(3291)	**IC** RV	*RV*	EH	
1207	(3328, 6422)	**V** BE		ZG	
1210	(3405, 6462)	**FS** E	*SR*	IS	
1211	(3305)	**PC** RA	*ST*	CS	
1212	(3427, 6453)	**V** RV	*RV*	EH	
1220	(3315, 6432)	**FS** E	*SR*	IS	
1221	(3371)	**IC** BE		ZG	
1254	(3391)	**BG** DR		BH	

KITCHEN WITH BAR

Mark 1. Built with no seats but three Pullman-style seats now fitted in bar area. B5 bogies. ETH 1.

Lot No. 30624 Cravens 1960–61. 41 t.

1566	**VN**	BE	*NB*	CP

KITCHEN BUFFET UNCLASSIFIED

Mark 1. Built with 23 loose chairs. All remaining vehicles refurbished with 23 fixed polypropylene chairs and fluorescent lighting. 1683/91/92/99 were further refurbished with 21 chairs, wheelchair space and carpets. ETH 2 (* 2X).

Now used on excursion trains with the seating area adapted to various uses including servery and food preparation areas, with some or all seating removed.

1651–99. Lot No. 30628 Pressed Steel 1960–61. Commonwealth bogies. 39 t.
1730. Lot No. 30512 BRCW 1960–61. B5 bogies. 37 t.

1651		**CC**	RV	*RV*	EH	1683		**RB**	RV	*RV*	EH
1657		**CH**	RV		ZG	1691		**CC**	RV	*RV*	EH
1659		**PC**	RA	*ST*	CS	1692		**CH**	RV		EH
1666	x	**M**	RP	*WC*	CS	1699		**RB**	RV		EH
1671	x*	**CH**	RV	*RV*	EH	1730	x	**M**	BK	*BK*	BT

BUFFET STANDARD

Mark 1. These vehicles are basically an open standard with two full window spaces removed to accommodate a buffet counter, and four seats removed to allow for a stock cupboard. All remaining vehicles now have fluorescent lighting. –/44 2T. Commonwealth bogies. ETH 3.

1861 has had its toilets replaced with store cupboards.

1813–32. Lot No. 30520 Wolverton 1960. 38 t.
1840. Lot No. 30507 Wolverton 1960. 37 t.
1859–63. Lot No. 30670 Wolverton 1961–62. 38 t.
1882. Lot No. 30702 Wolverton 1962. 38 t.

1813	x	**CH**	RV	*RV*	EH		1860	x	**M**	WC	*WC*	CS
1832	x	**CC**	RV	*RV*	EH		1861	x	**M**	WC	*WC*	CS
1840	v	**M**	WC	*WC*	CS		1863	x	**CH**	LS		CM
1859	x	**M**	BK	*BK*	BT		1882	x	**M**	WC	*WC*	CS

KITCHEN UNCLASSIFIED

Mark 1. These vehicles were built as Unclassified Restaurants. They were rebuilt with buffet counters and 23 fixed polypropylene chairs, then further refurbished by fitting fluorescent lighting. Further modified for use as servery vehicle with seating removed and kitchen extended. ETH 2X.

1953. Lot No. 30575 Swindon 1960. B4/B5 bogies. 36.5 t.
1961. Lot No. 30632 Swindon 1961. Commonwealth bogies. 39 t.

1953		**VN**	BE	*NB*	CP		1961	x	**M**	WC	*WC*	CS

HM THE QUEEN'S SALOON

Mark 3. Converted from an Open First built 1972. Consists of a lounge, bedroom and bathroom for HM The Queen, and a combined bedroom and bathroom for the Queen's dresser. One entrance vestibule has double doors. Air conditioned. BT10 bogies. ETH 9X.

Lot No. 30886 Wolverton 1977. 36 t.

2903	(11001)	**RP**	NR	*RP*		ZN

HRH THE DUKE OF EDINBURGH'S SALOON

Mark 3. Converted from an Open Standard built 1972. Consists of a combined lounge/dining room, a bedroom and a shower room for the Duke, a kitchen and a valet's bedroom and bathroom. Air conditioned. BT10 bogies. ETH 15X.

Lot No. 30887 Wolverton 1977. 36 t.

2904	(12001)	**RP**	NR	*RP*		ZN

ROYAL HOUSEHOLD SLEEPING CAR

Mark 3A. Built to similar specification as Sleeping Cars 10647–729. 12 sleeping compartments for use of Royal Household with a fixed lower berth and a hinged upper berth. 2T plus shower room. Air conditioned. BT10 bogies. ETH 11X.

Lot No. 31002 Derby/Wolverton 1985. 44 t.

2915   **RP** NR *RP*   ZN

HRH THE PRINCE OF WALES'S DINING CAR

Mark 3. Converted from HST TRUK built 1976. Large kitchen retained, but dining area modified for Royal use seating up to 14 at central table(s). Air conditioned. BT10 bogies. ETH 13X.

Lot No. 31059 Wolverton 1988. 43 t.

2916   (40512)   **RP** NR *RP*   ZN

ROYAL KITCHEN/HOUSEHOLD DINING CAR

Mark 3. Converted from HST TRUK built 1977. Large kitchen retained and dining area slightly modified with seating for 22 Royal Household members. Air conditioned. BT10 bogies. ETH 13X.

Lot No. 31084 Wolverton 1990. 43 t.

2917   (40514)   **RP** NR *RP*   ZN

ROYAL HOUSEHOLD CARS

Mark 3. Converted from HST TRUKs built 1976/77. Air conditioned. BT10 bogies. ETH 10X.

Lot Nos. 31083 (* 31085) Wolverton 1989. 41.05 t.

2918   (40515)   **RP** NR   ZN
2919   (40518) *   **RP** NR   ZN

ROYAL HOUSEHOLD COUCHETTES

Mark 2B. Converted from Corridor Brake First built 1969. Consists of luggage accommodation, guard's compartment, workshop area, 350 kW diesel generator and staff sleeping accommodation. B5 bogies. ETH 2X.

Lot No. 31044 Wolverton 1986. 48 t.

2920   (14109, 17109) **RP** NR *RP*   ZN

Mark 2B. Converted from Corridor Brake First built 1969. Consists of luggage accommodation, kitchen, brake control equipment and staff accommodation. B5 bogies. ETH 7X.

Lot No. 31086 Wolverton 1990. 41.5 t.

2921   (14107, 17107)   **RP**   NR   *RP*   ZN

HRH THE PRINCE OF WALES'S SLEEPING CAR

Mark 3B. Air conditioned. BT10 bogies. ETH 7X.

Lot No. 31035 Derby/Wolverton 1987.

2922   **RP** NR *RP*   ZN

ROYAL SALOON

Mark 3B. Air conditioned. BT10 bogies. ETH 6X.

Lot No. 31036 Derby/Wolverton 1987.

2923   **RP** NR *RP*   ZN

OPEN FIRST

Mark 1. 42/– 2T. ETH 3. Many now fitted with table lamps.

3058 was numbered DB 975313, 3068 was numbered DB 975606 and 3093 was numbered DB 977594 for a time when in departmental service for BR.

3058–69. Lot No. 30169 Doncaster 1955. B4 bogies. 33 t (* Commonwealth bogies 35 t).
3093. Lot No. 30472 BRCW 1959. B4 bogies. 33 t.
3096–3100. Lot No. 30576 BRCW 1959. B4 bogies. 33 t.

3058	*x **M**	WC *WC*	CS	FLORENCE	3096	x **M**	BK *BK*	BT
3066	**CC**	RV *RV*	EH		3097	**CC**	RV *RV*	EH
3068	**CC**	RV *RV*	EH		3098	x **CH**	RV *RV*	EH
3069	**CC**	RV *RV*	EH		3100	x **CH**	RV *RV*	EH
3093	x **M**	WC *WC*	CS	FLORENCE				

Later design with fluorescent lighting, aluminium window frames and Commonwealth bogies.

3128/36/41/43/44/46/47/48 were renumbered 1058/60/63/65/66/68/69/70 when reclassified Restaurant Open First, then 3600/05/08/09/02/06/04/10 when declassified to Open Standard, but have since regained their original numbers. 3136 was numbered DB 977970 for a time when in use with Serco Railtest as a Brake Force Runner.

3105 has had its luggage racks removed and has tungsten lighting.

3105–28. Lot No. 30697 Swindon 1962–63. 36 t.
3130–50. Lot No. 30717 Swindon 1963. 36 t.

3105	x	**M**	WC	*WC*	CS
3106	x	**M**	WC	*WC*	CS
3107	x	**CH**	RV	*RV*	EH
3110	x	**CH**	RV	*RV*	EH
3112	x	**CH**	RV	*RV*	EH
3113	x	**M**	WC	*WC*	CS
3115	x	**M**	BK	*BK*	BT
3117	x	**M**	WC	*WC*	CS
3119		**CC**	RV	*RV*	EH
3120		**CC**	RV	*RV*	EH
3121		**CH**	RV	*RV*	EH
3122	x	**CH**	RV	*RV*	EH
3123		**CC**	RV	*RV*	EH
3125	x	**CH**	LS	*RV*	EH

3128	x	**M**	WC	*WC*	CS
3130	x	**M**	WC	*WC*	CS
3133	x	**M**	RV		EH
3136	x	**M**	WC	*WC*	CS
3140	x	**CH**	RV	*RV*	EH
3141		**M**	RV		CD
3143	x	**M**	WC	*WC*	CS
3144	x	**M**	RV		BQ
3146		**M**	RV		EH
3147		**CH**	RV	*RV*	EH
3148		**M**	LS		CL
3149		**CC**	RV	*RV*	EH
3150		**M**	BK	*BK*	BT

Names:

3105	JULIA	3128	VICTORIA
3106	ALEXANDRA	3130	PAMELA
3113	JESSICA	3136	DIANA
3117	CHRISTINA	3143	PATRICIA

OPEN FIRST

Mark 2D. Air conditioned. Stones equipment. 42/– 2T. B4 bogies. ETH 5.

† Interior modified to Pullman Car standards with new seating, new
   panelling, tungsten lighting and table lights for the Belmond Northern Belle.

Lot No. 30821 Derby 1971–72. 34 t.

3174	†	**VN**	BE	*NB*	CP	GLAMIS
3182	†	**VN**	BE	*NB*	CP	WARWICK
3188		**PC**	RA	*ST*	CS	CADAIR IDRIS

OPEN FIRST

Mark 2E. Air conditioned. Stones equipment. 42/– 2T (p 36/– 2T). B4 bogies.
ETH 5.

r Refurbished with new seats.
† Interior modified to Pullman Car standards with new seating, new
   panelling, tungsten lighting and table lights for the Belmond Northern Belle.

Lot No. 30843 Derby 1972–73. 32.5 t. († 35.8 t).

3223		**RV**	RA		BO	DIAMOND
3231	p	**PC**	RA	*ST*	CS	BEN CRUACHAN
3232	dr	**BG**	BE		CD	
3240		**RV**	RA		BO	SAPPHIRE
3247	†	**VN**	BE	*NB*	CP	CHATSWORTH
3267	†	**VN**	BE	*NB*	CP	BELVOIR
3273	†	**VN**	BE	*NB*	CP	ALNWICK
3275	†	**VN**	BE	*NB*	CP	HARLECH

OPEN FIRST

Mark 2F. Air conditioned. 3277–3318/3358–79 have Stones equipment, others have Temperature Ltd. All refurbished in the 1980s with power-operated vestibule doors, new panels and new seat trim. 42/– 2T. B4 bogies. d. ETH 5X.

r   Further refurbished with table lamps and modified seats with burgundy seat trim.

u   Fitted with power supply for Mark 1 Kitchen Buffet Unclassified.

3277–3318. Lot No. 30845 Derby 1973. 33.5 t.
3325–3426. Lot No. 30859 Derby 1973–74. 33.5 t.
3431–38. Lot No. 30873 Derby 1974–75. 33.5 t.

3277		**AR**	RV	*RV*	EH		3356	r	**RV**	RV	*RV*	EH
3278	r	**BP**	RV	*RV*	EH		3358		**M**	DB		MH
3279	u	**M**	DB		MH		3359	r	**M**	WC	*WC*	CS
3292		**M**	DB		MH		3360	r	**PC**	WC	*WC*	CS
3295		**AR**	RV	*RV*	EH		3362	r	**PC**	WC	*WC*	CS
3304	r	**V**	RV	*RV*	EH		3364	r	**RV**	RV	*RV*	EH
3312		**PC**	RA	*ST*	CS		3366	r	**BG**	DR		BH
3313	r	**M**	WC	*WC*	CS		3374		**BG**	DR		BH
3314	r	**V**	RV	*RV*	EH		3379	u	**AR**	RV	*RV*	EH
3318		**M**	DB		MH		3384	r	**RV**	RV	*RV*	EH
3325	r	**V**	RV	*RV*	EH		3386	r	**V**	RV	*RV*	EH
3326	r	**M**	WC	*WC*	CS		3388		**M**	DB		Fawley
3330	r	**RV**	RV	*RV*	EH		3390	r	**RV**	RV	*RV*	EH
3331		**M**	DB		MH		3392	r	**M**	WC	*WC*	CS
3333	r	**V**	RV	*RV*	EH		3395	r	**M**	WC	*WC*	CS
3334		**AR**	RV	*RV*	EH		3397	r	**RV**	RV	*RV*	EH
3336	u	**AR**	RV	*RV*	EH		3399	u	**M**	DB		Fawley
3340	r	**V**	RV	*RV*	EH		3400		**M**	DB		MH
3344	r	**V**	RV	*RV*	EH		3417		**AR**	RV	*RV*	EH
3345	r	**V**	RV	*RV*	EH		3424		**M**	DB		MH
3348	r	**RV**	RV	*RV*	EH		3426	r	**RV**	RV	*RV*	EH
3350	r	**M**	WC	*WC*	CS		3431	r	**M**	WC	*WC*	CS
3351		**CH**	VT	*WC*	CS		3438	r	**PC**	RA	*ST*	CS
3352	r	**M**	WC	*WC*	CS							

Names:

3312	HELVELLYN		3384	DICKENS
3330	BRUNEL		3390	CONSTABLE
3348	GAINSBOROUGH		3397	WORDSWORTH
3356	TENNYSON		3426	ELGAR
3364	SHAKESPEARE		3438	BEN LOMOND

OPEN STANDARD

Mark 1. This coach has narrower seats than later vehicles. Built with BR Mark 1 bogies. –/64 2T. ETH 4.

Lot No. 30079 York 1953. Commonwealth bogies. 36 t.

| 3766 | x | **M** | WC | *WC* | | CS |

OPEN STANDARD

Mark 1. These vehicles are a development of the above with fluorescent lighting and modified design of seat headrest. Built with BR Mark 1 bogies. –/64 2T. ETH 4.

4831–36. Lot No. 30506 Wolverton 1959. Commonwealth bogies. 33 t.
4856. Lot No. 30525 Wolverton 1959–60. B4 bogies. 33 t.

| 4831 | x | **M** | BK | *BK* | BT | | 4836 | x | **M** | BK | *BK* | BT |
| 4832 | x | **M** | BK | *BK* | BT | | 4856 | x | **M** | BK | *BK* | BT |

OPEN STANDARD

Mark 1. Later vehicles built with Commonwealth bogies. –/64 2T. ETH 4:

4905. Lot No. 30646 Wolverton 1961. 36 t.
4927–5044. Lot No. 30690 Wolverton 1961–62. Aluminium window frames. 37 t.

4905	x	**M**	WC	*WC*	CS		4994	x	**M**	WC	*WC*	CS
4927		**CC**	RV	*RV*	EH		4998		**CH**	RV	*RV*	EH
4931	v	**M**	WC	*WC*	CS		5007		**G**	RV		EH
4940	x	**M**	WC	*WC*	CS		5008	x	**M**	RV		EH
4946	x	**CH**	RV	*RV*	EH		5009	x	**CH**	RV		EH
4949	x	**CH**	RV	*RV*	EH		5027		**G**	RV		EH
4951	x	**M**	WC	*WC*	CS		5028	x	**CC**	BK	*BK*	BT
4954	v	**M**	WC	*WC*	CS		5032	x	**M**	WC	*WC*	CS
4959		**CH**	RV	*RV*	EH		5033	x	**M**	WC	*WC*	CS
4960	x	**M**	WC	*WC*	CS		5035	x	**M**	WC	*WC*	CS
4973	x	**M**	WC	*WC*	CS		5040	x	**CH**	RV	*RV*	EH
4984	x	**M**	WC	*WC*	CS		5044	x	**M**	WC	*WC*	CS
4991		**CH**	RV	*RV*	EH							

OPEN STANDARD

Mark 2. Pressure ventilated. –/64 2T. B4 bogies. ETH 4.

Lot No. 30751 Derby 1965–67. 32 t.

5157	v	**CH**	VT	*VT*	TM		5200	v	**M**	WC	*WC*	CS
5171	v	**M**	WC	*WC*	CS		5212	v	**CH**	VT	*VT*	TM
5177	v	**CH**	VT	*VT*	TM		5216	v	**M**	WC	*WC*	CS
5191	v	**CH**	VT	*VT*	TM		5222	v	**M**	WC	*WC*	CS
5198	v	**CH**	VT	*VT*	TM							

OPEN STANDARD

Mark 2. Pressure ventilated. –/48 2T. B4 bogies. ETH 4.

Lot No. 30752 Derby 1966. 32 t.

5229		**M**	WC	*WC*	CS	5239		**M**	WC	*WC*	CS
5236	v	**M**	WC	*WC*	CS	5249	v	**M**	WC	*WC*	CS
5237	v	**M**	WC	*WC*	CS						

OPEN STANDARD

Mark 2A. Pressure ventilated. –/64 2T (w –/62 2T). B4 bogies. ETH 4.

f Facelifted vehicles.

5276–5341. Lot No. 30776 Derby 1967–68. 32 t.
5366–5419. Lot No. 30787 Derby 1968. 32 t.

5276	f	**RV**	RV		BQ	5341	f	**CC**	RV	*RV*	EH
5278		**M**	WC	*WC*	CS	5366	f	**CH**	RV	*RV*	EH
5292	f	**CC**	RV	*RV*	EH	5419	w	**M**	WC	*WC*	CS

OPEN STANDARD

Mark 2D. Air conditioned. Stones equipment. Refurbished with new seats and end luggage stacks. –/58 2T. B4 bogies. d. ETH 5.

Lot No. 30822 Derby 1971. 33 t.

5631	**M**	DB	*SR*	TO	5700	**FP**	DR		BH
5632	**M**	DB	*SR*	TO	5710	**FP**	DR		BH
5657	**M**	DB	*SR*	TO					

OPEN STANDARD

Mark 2E. Air conditioned. Stones equipment. –/64 2T. B4 bogies. d. ETH 5.

r Refurbished with new interior panelling.
s Refurbished with new interior panelling, modified design of seat headrest and centre luggage stack. –/60 2T.

5787–97. Lot No. 30837 Derby 1972. 33.5 t.
5810. Lot No. 30844 Derby 1972–73. 33.5 t.

5787	s	**V**	DR		BH	5810	s	**DS**	DR	*DR*	KM
5797	r★	**IC**	RA		BO						

OPEN STANDARD

Mark 2F. Air conditioned. Temperature Ltd equipment. InterCity 70 seats. All were refurbished in the 1980s with power-operated vestibule doors, new panels and seat trim. –/64 2T. B4 bogies. d. ETH 5X.

* Early Mark 2 style seats. These vehicles have undergone a second refurbishment with carpets and new seat trim.
q Fitted with two wheelchair spaces. –/60 2T 2W.
s Fitted with centre luggage stack. –/60 2T.
t Fitted with centre luggage stack and wheelchair space. –/58 2T 1W.

5910–55. Lot No. 30846 Derby 1973. 33 t.
5959–6158. Lot No. 30860 Derby 1973–74. 33 t.
6173–83. Lot No. 30874 Derby 1974–75. 33 t.

No.	Note				Depot
5910	q	V	RV	RV	EH
5912		PC	RA	ST	CS
5919	s pt	DS	DR	DR	KM
5921		AR	RV	RV	EH
5922		M	DB		BS
5924		M	DB		BS
5928		CH	VT	WC	CS
5929		AR	RV	RV	EH
5937		V	RV	RV	EH
5945		V	RV	RV	EH
5950		AR	RV	RV	EH
5952		V	RV	RV	EH
5954		M	DB	SR	TO
5955		V	RV	RV	EH
5959	n	M	DB		BS
5961	s pt	V	RV	RV	EH
5964		AR	RV	RV	EH
5965	t	AW	RV	RV	EH
5971		DS	DR	DR	KM
5976	t	AW	RV	RV	EH
5985		AR	RV	RV	EH
5987		V	RV	RV	EH
5991		PC	RA	ST	CS
5995		DS	DR	GA	KM
5998		AR	RV	RV	EH
6000	t	M	WC	WC	CS
6001		DS	DR	GA	KM
6006		AR	RV	RV	EH
6008		DS	DR	GA	KM
6012		M	WC	WC	CS
6021		M	WC	WC	CS
6022	s	M	WC	WC	CS
6024	s	V	RV	RV	EH
6027	q	V	RV	RV	EH
6036	*	M	DB		BS
6042		AR	RV	RV	EH
6046		DS	DR	DR	KM
6051		V	RV	RV	EH
6054		V	RV	RV	EH
6064	s	DS	DR	DR	KM
6067	s pt	V	RV	RV	EH
6103		M	WC	WC	CS
6110		M	DB	SR	TO
6115	s	M	WC	WC	CS
6117	★	DS	DR	DR	KM
6122	★	DS	DR	DR	KM
6137	s pt	AW	RV	RV	EH
6139	n*	M	DB		MH
6141	q	V	RV	RV	EH
6152	*	M	DB		BS
6158		V	RV	RV	EH
6173	★	DS	DR	DR	KM
6176	t	V	RV	RV	EH
6177	s	V	RV	RV	EH
6183	s	AW	RV	RV	EH

BRAKE GENERATOR VAN

Mark 1. Renumbered 1989 from BR departmental series. Converted from Gangwayed Brake Van in 1973 to three-phase supply brake generator van for use with HST trailers. Modified 1999 for use with loco-hauled stock. B5 bogies.

Lot No. 30400 Pressed Steel 1958.

6310	(81448, 975325)	**CH**	RV	*RV*	EH

GENERATOR VAN

Mark 1. Converted from Gangwayed Brake Vans in 1992. B4 bogies. ETH 75.

6311. Lot No. 30162 Pressed Steel 1958. 37.25 t.
6312. Lot No. 30224 Cravens 1956. 37.25 t.
6313. Lot No. 30484 Pressed Steel 1958. 37.25 t.

6311	(80903, 92911)	**B**	DB		TO
6312	(81023, 92925)	**M**	WC	*WC*	CS
6313	(81553, 92167)	**PC**	BE	*BP*	SL

BUFFET STANDARD

Mark 2C. Converted from Open Standard by removal of one seating bay and replacing this by a counter with a space for a trolley, now replaced with a more substantial buffet. Adjacent toilet removed and converted to steward's washing area/store. Pressure ventilated. –/55 1T. B4 bogies. ETH 4.

Lot No. 30795 Derby 1969–70. 32.5 t.

6528	(5592)	**M**	WC	*WC*	CS

SLEEPER RECEPTION CAR

Mark 2F. Converted from Open First. These vehicles consist of pantry, microwave cooking facilities, seating area for passengers (with loose chairs, staff toilet plus two bars). Now refurbished again with new "sofa" seating as well as the loose chairs. Converted at RTC, Derby (6700), Ilford (6701–05) and Derby (6706–08). Air conditioned. 6700/01/03/05–08 have Stones equipment and 6702/04 have Temperature Ltd equipment. The number of seats per coach can vary but typically is 25/– 1T (12 seats as "sofa" seating and 13 loose chairs). B4 bogies. d. ETH 5X.

6700–02/04/08. Lot No. 30859 Derby 1973–74. 33.5 t.
6703/05–07. Lot No. 30845 Derby 1973. 33.5 t.

6700	(3347)	**FS**	E	*SR*	IS
6701	(3346)	**FS**	E	*SR*	IS
6702	(3421)	**FS**	E	*SR*	IS
6703	(3308)	**FS**	E	*SR*	IS
6704	(3341)	**FS**	E	*SR*	IS

6705	(3310, 6430)	**FS**	E	*SR*		IS
6706	(3283, 6421)	**FS**	E	*SR*		IS
6707	(3276, 6418)	**FS**	E	*SR*		IS
6708	(3370)	**FS**	E	*SR*		IS

BUFFET FIRST

Mark 2D. Converted from Buffet Standard by the removal of another seating bay and fitting a more substantial buffet counter with boiler and microwave oven. Now converted to First Class with new seating and end luggage stacks. Air conditioned. Stones equipment. 30/– 1T. B4 bogies. d. ETH 5.

Lot No. 30822 Derby 1971. 33 t.

6720	(5622, 6652)	**M**	DB		FA
6722	(5736, 6661)	**FP**	E		LM
6723	(5641, 6662)	**M**	WC		CS
6724	(5721, 6665)	**M**	WC	*WC*	CS

OPEN BRAKE STANDARD WITH TROLLEY SPACE

Mark 2. This vehicle uses the same body shell as the Mark 2 Corridor Brake First. Converted from Open Brake Standard by removal of one seating bay and replacing this with a counter with a space for a trolley. Adjacent toilet removed and converted to a steward's washing area/store. –/23. B4 bogies. ETH 4.

Lot No. 30757 Derby 1966. 31 t.

9101	(9398)	v	**CH**	VT	*VT*		TM

OPEN BRAKE STANDARD

Mark 2. These vehicles use the same body shell as the Mark 2 Corridor Brake First and have First Class seat spacing and wider tables. Pressure ventilated. –/31 1T. B4 bogies. ETH 4.

9104 was originally numbered 9401. It was renumbered when converted to Open Brake Standard with trolley space. Now returned to original layout.

Lot No. 30757 Derby 1966. 31.5 t.

9104	v	**M**	WC	*WC*	CS		9392	v	**M**	WC	*WC*	CS
9391		**M**	WC	*WC*	CS							

OPEN BRAKE STANDARD

Mark 2C. Pressure ventilated. –/31 1T. B4 bogies. d. ETH 4.

Lot No. 30798 Derby 1970. 32 t.

9440		**M**	WC	SH

OPEN BRAKE STANDARD

Mark 2D. Air conditioned. Stones Equipment. All now refurbished with new seating –/22 1TD. B4 bogies. d. pg. ETH 5.

Lot No. 30824 Derby 1971. 33 t.

9488		**FP**	DR		BH	9494	**M** DB	MH
9493		**M**	WC	*WC*	CS			

OPEN BRAKE STANDARD

Mark 2E. Air conditioned. Stones Equipment. –/32 1T. B4 bogies. d. pg. ETH 5.

Lot No. 30838 Derby 1972. 33 t.

Non-standard livery: 9502 Pullman umber & cream.

r Refurbished with new interior panelling.
s Refurbished with modified design of seat headrest and new interior panelling.

9496	r	**IC**	VT	*WC*	CS	9507	s	**V**	RV	*SR*	EH
9502	s	**0**	BE	*BP*	SL	9508	s	**BG**	DR		BH
9504	s	**V**	RV	*RV*	EH	9509	s	**AV**	RV		CD
9506	s★	**BG**	DR		MH						

OPEN BRAKE STANDARD

Mark 2F. Air conditioned. Temperature Ltd equipment. All were refurbished in the 1980s with power-operated vestibule doors, new panels and seat trim. All now further refurbished with carpets. –/32 1T. B4 bogies. d. pg. ETH 5X.

9537 has had all its seats removed and is used to carry luggage.

Advertising livery: 9537 CruiseSaver Express (dark blue).

Lot No. 30861 Derby 1974. 34 t.

9520	n	**AR**	RV	*RV*	EH	9527	n	**AR**	RV	*RV*	EH
9521	★	**AW**	RV	*RV*	EH	9529	n	**M**	DB		BS
9522		**M**	DB	*SR*	TO	9531		**M**	DB		BS
9525		**DS**	DR	*GA*	KM	9537	n	**AL**	RV		EH
9526	n★	**IC**	RV	*RV*	EH	9539		**AW**	RV	*RV*	EH

DRIVING OPEN BRAKE STANDARD

Mark 2F. Air conditioned. Temperature Ltd equipment. Push & pull (tdm system). Converted from Open Brake Standard, these vehicles originally had half cabs at the brake end. They have since been refurbished and have had their cabs widened and the cab-end gangways removed. Five vehicles (9701–03/08/14) have been converted for use in Network Rail test trains and can be found in the Service Stock section of this book. –/30 1W 1T. B4 bogies. d. pg. Cowcatchers. ETH 5X.

9704–10. Lot No. 30861 Derby 1974. Converted Glasgow 1979. Disc brakes. 34 t.
9711/13. Lot No. 30861 Derby 1974. Converted Glasgow 1985. 34 t.

9704	(9512)	**AR**	BA	ZG		9710	(9518)	**1**	BA	ZG
9705	(9519)	**AR**	DR	ZA		9711	(9532)	**AR**	VT	TM
9707	(9511)	**AR**	DR	ZA		9713	(9535)	**AR**	NR	ZA
9709	(9515)	**AR**	BA	ZG						

OPEN BRAKE UNCLASSIFIED

Mark 2E. Converted from Open Standard with new seating for use on Anglo-Scottish overnight services by Railcare, Wolverton. Air conditioned. Stones equipment. –/31 2T. B4 bogies. d. ETH 4X.

9801–03. Lot No. 30837 Derby 1972. 33.5 t.
9804–10. Lot No. 30844 Derby 1972–73. 33.5 t.

9800	(5751)	**FS**	E	*SR*	IS	9806	(5840)	**FS**	E	*SR*	IS
9801	(5760)	**FS**	E	*SR*	IS	9807	(5851)	**FS**	E	*SR*	IS
9802	(5772)	**FS**	E	*SR*	IS	9808	(5871)	**FS**	E	*SR*	IS
9803	(5799)	**FS**	E	*SR*	IS	9809	(5890)	**FS**	E	*SR*	IS
9804	(5826)	**FS**	E	*SR*	IS	9810	(5892)	**FS**	E	*SR*	IS
9805	(5833)	**FS**	E	*SR*	IS						

KITCHEN BUFFET FIRST

Mark 3A. Air conditioned. Converted from HST catering vehicles and Mark 3 Open Firsts. Refurbished with table lamps and burgundy seat trim (except *). 18/– plus two seats for staff use (* 24/–, † 24/–, § 24/–, t 23/– 1W). BT10 bogies. d. ETH 14X.

§ First Great Western Sleeper "day coaches" that have been fitted with former HST First Class seats to a 2+1 layout.

Non-standard livery: 10211 EWS dark maroon.

10200–211. Lot No. 30884 Derby 1977. 39.8 t.
10212–229. Lot No. 30878 Derby 1975–76. 39.8 t.
10232–259. Lot No. 30890 Derby 1979. 39.8 t.

10200	(40519)	*	**GA**	P		*GA*	NC	10228	(11035)	*	**1**	P		*GA*	NC
10202	(40504)	†	**BG**	AV			LM	10229	(11059)	*	**GA**	P		*GA*	NC
10203	(40506)	*	**GA**	P		*GA*	NC	10232	(10027)	§	**FD**	P		*GW*	PZ
10206	(40507)		**V**	P			LM	10233	(10013)		**V**	AV			LM
10211	(40510)		**0**	DB	*DB*		TO	10235	(10015)	†	**BG**	AV			LM
10212	(11049)		**VT**	P		*VW*	WB	10237	(10022)		**DR**	AV			LM
10214	(11034)	*	**1**	P		*GA*	NC	10241	(10009)	*	**1**	P			IL
10215	(11032)		**BG**	AV			LM	10242	(10002)		**BG**	AV			LM
10216	(11041)	*	**GA**	P		*GA*	NC	10246	(10014)	†	**BG**	AV	*AW*	CF
10217	(11051)		**VT**	P			LM	10247	(10011)	*	**NX**	P		*GA*	NC
10219	(11047)	§	**FD**	P		*GW*	PZ	10249	(10012)		**AW**	AV	*AW*	CF
10222	(11063)		**BG**	AV			LM	10250	(10020)		**V**	AV			LM
10223	(11043)	*	**1**	P		*GA*	NC	10253	(10026)		**V**	P			LM
10225	(11014)	§	**FD**	P		*GW*	PZ	10257	(10007)	†	**BG**	AV			LM
10226	(11015)		**V**	AV			LM	10259	(10025)	t	**AW**	AV	*AW*	CF

KITCHEN BUFFET FIRST

Mark 3A. Air conditioned. Rebuilt for Chiltern Railways 2011–12 and fitted with sliding plug doors. Interiors originally refurbished for Wrexham & Shropshire with Primarius seating, a new kitchen area and universal-access toilet. 30/– 1TD 1W. BT10 bogies. ETH 14X.

10271/273/274. Lot No. 30890 Derby 1979. 41.3 t.
10272. Lot No. 30884 Derby 1977. 41.3 t.

10271	(10018, 10236)	**CM**	AV	*CR*	AL
10272	(40517, 10208)	**CM**	AV	*CR*	AL
10273	(10021, 10230)	**CM**	AV	*CR*	AL
10274	(10010, 10255)	**CM**	AV	*CR*	AL

KITCHEN BUFFET STANDARD

Mark 4. Air conditioned. Rebuilt from First to Standard Class with bar adjacent to seating area instead of adjacent to end of coach. –/30 1T. BT41 bogies. ETH 6X.

Advertising livery: 10312 Sky 1 (blue).

Lot No. 31045 Metro-Cammell 1989–92. 43.2 t.

10300	**EC**	E	*EC*	BN	10317	**EC**	E	*EC*	BN
10301	**EC**	E	*EC*	BN	10318	**EC**	E	*EC*	BN
10302	**EC**	E	*EC*	BN	10319	**EC**	E	*EC*	BN
10303	**EC**	E	*EC*	BN	10320	**EC**	E	*EC*	BN
10304	**EC**	E	*EC*	BN	10321	**EC**	E	*EC*	BN
10305	**EC**	E	*EC*	BN	10323	**EC**	E	*EC*	BN
10306	**EC**	E	*EC*	BN	10324	**EC**	E	*EC*	BN
10307	**EC**	E	*EC*	BN	10325	**EC**	E	*EC*	BN
10308	**EC**	E	*EC*	BN	10326	**EC**	E	*EC*	BN
10309	**EC**	E	*EC*	BN	10328	**EC**	E	*EC*	BN
10310	**EC**	E	*EC*	BN	10329	**EC**	E	*EC*	BN
10311	**EC**	E	*EC*	BN	10330	**EC**	E	*EC*	BN
10312	**AL**	E	*EC*	BN	10331	**EC**	E	*EC*	BN
10313	**EC**	E	*EC*	BN	10332	**EC**	E	*EC*	BN
10315	**EC**	E	*EC*	BN	10333	**EC**	E	*EC*	BN

BUFFET STANDARD

Mark 3A. Air conditioned. Converted from Mark 3 Open Standards at Derby 2006. –/52 1T (including 6 Compin Pegasus seats for "priority" use). BT10 bogies. d. ETH 13X.

Lot No. 30877 Derby 1975–77. 37.8 t.

10401	(12168)	**NX**	P	*GA*	NC	10404	(12068)	**1**	P	*GA*	NC
10402	(12010)	**1**	P	*GA*	NC	10405	(12157)	**1**	P	*GA*	NC
10403	(12135)	**1**	P	*EA*	NC	10406	(12020)	**1**	P	*GA*	NC

SLEEPING CAR WITH PANTRY

Mark 3A. Air conditioned. Retention toilets. 12 compartments with a fixed lower berth and a hinged upper berth, plus an attendants compartment. 2T. BT10 bogies. d. ETH 7X.

Non-standard livery: 10546 EWS dark maroon.

Lot No. 30960 Derby 1981–83. 41 t.

10501	**FS**	P	*SR*	IS		10553	**FS**	P	*SR*	IS
10502	**FS**	P	*SR*	IS		10561	**FS**	P	*SR*	IS
10504	**FS**	P	*SR*	IS		10562	**FS**	P	*SR*	IS
10506	**FS**	P	*SR*	IS		10563	**FD**	P	*GW*	PZ
10507	**FS**	P	*SR*	IS		10565	**FS**	P	*SR*	IS
10508	**FS**	P	*SR*	IS		10580	**FS**	P	*SR*	IS
10513	**FS**	P	*SR*	IS		10584	**FD**	P	*GW*	PZ
10516	**FS**	P	*SR*	IS		10589	**FD**	P	*GW*	PZ
10519	**FS**	P	*SR*	IS		10590	**FD**	P	*GW*	PZ
10520	**FS**	P	*SR*	IS		10594	**FD**	P	*GW*	PZ
10522	**FS**	P	*SR*	IS		10596	**U**	P		ZH
10523	**FS**	P	*SR*	IS		10597	**FS**	P	*SR*	IS
10526	**FS**	P	*SR*	IS		10598	**FS**	P	*SR*	IS
10527	**FS**	P	*SR*	IS		10600	**FS**	P	*SR*	IS
10529	**FS**	P	*SR*	IS		10601	**FD**	P	*GW*	PZ
10531	**FS**	P	*SR*	IS		10605	**FS**	P	*SR*	IS
10532	**FD**	P	*GW*	PZ		10607	**FS**	P	*SR*	IS
10534	**FD**	P	*GW*	PZ		10610	**FS**	P	*SR*	IS
10542	**FS**	P	*SR*	IS		10612	**FD**	P	*GW*	PZ
10543	**FS**	P	*SR*	IS		10613	**FS**	P	*SR*	IS
10544	**FS**	P	*SR*	IS		10614	**FS**	P	*SR*	IS
10546	**0**	DB	*DB*	TO		10616	**FD**	P	*GW*	PZ
10548	**FS**	P	*SR*	IS		10617	**FS**	P	*SR*	IS
10551	**FS**	P	*SR*	IS						

SLEEPING CAR

Mark 3A. Air conditioned. Retention toilets. 13 compartments with a fixed lower berth and a hinged upper berth (* 11 compartments with a fixed lower berth and a hinged upper berth + one compartment for a disabled person. 1TD). 2T. BT10 bogies. ETH 6X.

10734 was originally 2914 and used as a Royal Train staff sleeping car. It has 12 berths and a shower room and is ETH 11X.

10648–729. Lot No. 30961 Derby 1980–84. 43.5 t.
10734. Lot No. 31002 Derby/Wolverton 1985. 42.5 t.

10648	d*	**FS**	P	*SR*	IS	10688	d	**FS**	P	*SR*	IS
10650	d*	**FS**	P	*SR*	IS	10689	d*	**FS**	P	*SR*	IS
10666	d*	**FS**	P	*SR*	IS	10690	d	**FS**	P	*SR*	IS
10675	d	**FS**	P	*SR*	IS	10693	d	**FS**	P	*SR*	IS
10680	d*	**FS**	P	*SR*	IS	10699	d*	**FS**	P	*SR*	IS
10683	d	**FS**	P	*SR*	IS	10703	d	**FS**	P	*SR*	IS

10706	d*	**FS**	P	*SR*	IS		10722	d*	**FS**	P	*SR*	IS
10714	d*	**FS**	P	*SR*	IS		10723	d*	**FS**	P	*SR*	IS
10718	d*	**FS**	P	*SR*	IS		10729		**VN**	BE	*NB*	CP
10719	d*	**FS**	P	*SR*	IS		10734		**VN**	BE	*NB*	CP

Names:

10729	CREWE	10734	BALMORAL

OPEN FIRST

Mark 3A. Air conditioned. All refurbished with table lamps and new seat cushions and trim. 48/– 2T (* 48/– 1T 1TD, † 47/– 2T 1W). BT10 bogies. d. ETH 6X.

§ 11029 has been reseated with Standard Class seats: –/68 2T 2W.

11006/007 were open composites 11906/907 for a time.

Non-standard livery: 11039 EWS dark maroon.

Lot No. 30878 Derby 1975–76. 34.3 t.

11006		**V**	DR		BH		11028		**V**	AV		ZB
11007		**VT**	P	*VW*	WB		11029	§	**BG**	AV	*CR*	AL
11011	*	**V**	DR		BH		11031	†	**BG**	AV	*CR*	AL
11018		**VT**	P	*VW*	WB		11033		**DR**	AV		LM
11021		**V**	P		LM		11039		**0**	DB	*DB*	TO
11026		**V**	P		LM		11048		**VT**	P	*VW*	WB

OPEN FIRST

Mark 3B. Air conditioned. InterCity 80 seats. All refurbished with table lamps and new seat cushions and trim. 48/– 2T. BT10 bogies. d. ETH 6X.

† Greater Anglia vehicles fitted with disabled toilet and reduced seating including three Compin "Pegasus" seats of the same type as used in Standard Class (but regarded as First Class!). 34/3 1T 1TD 2W.

Lot No. 30982 Derby 1985. 36.5 t.

11066		**1**	P	*GA*	NC		11085	†	**1**	P	*GA*	NC
11067		**1**	P	*GA*	NC		11087	†	**GA**	P	*GA*	NC
11068		**1**	P	*GA*	NC		11088	†	**GA**	P	*GA*	NC
11069		**GA**	P	*GA*	NC		11090	†	**1**	P	*GA*	NC
11070		**1**	P	*GA*	NC		11091		**GA**	P	*GA*	NC
11072		**1**	P	*GA*	NC		11092	†	**1**	P	*GA*	NC
11073		**1**	P	*GA*	NC		11093	†	**GA**	P	*GA*	NC
11074		**V**	P		LM		11094	†	**1**	P	*GA*	NC
11075		**GA**	P	*GA*	NC		11095	†	**1**	P	*GA*	NC
11076		**1**	P	*GA*	NC		11096	†	**GA**	P	*GA*	NC
11077		**1**	P	*GA*	NC		11097		**V**	AV		LM
11078	†	**1**	P	*GA*	NC		11098	†	**1**	P	*GA*	NC
11079		**V**	AV		LM		11099	†	**1**	P	*GA*	NC
11080		**1**	P	*GA*	NC		11100	†	**1**	P	*GA*	NC
11081		**1**	P	*GA*	NC		11101	†	**1**	P	*GA*	NC
11082		**GA**	P	*GA*	NC							

OPEN FIRST

Mark 4. Air conditioned. Rebuilt with new interior by Bombardier Wakefield 2003–05 (some converted from Standard Class vehicles) 41/– 1T (plus 2 seats for staff use). BT41 bogies. ETH 6X.

Advertising livery: 11294 Sky 1 (blue).

11201–11273. Lot No. 31046 Metro-Cammell 1989–92. 41.3 t.
11277–11299. Lot No. 31049 Metro-Cammell 1989–92. 41.3 t.

11201		**EC**	E	*EC*	BN	11284 (12487)	**EC**	E	*EC* BN
11219		**EC**	E	*EC*	BN	11285 (12537)	**EC**	E	*EC* BN
11229		**EC**	E	*EC*	BN	11286 (12482)	**EC**	E	*EC* BN
11237		**EC**	E	*EC*	BN	11287 (12527)	**EC**	E	*EC* BN
11241		**EC**	E	*EC*	BN	11288 (12517)	**EC**	E	*EC* BN
11244		**EC**	E	*EC*	BN	11289 (12528)	**EC**	E	*EC* BN
11273		**EC**	E	*EC*	BN	11290 (12530)	**EC**	E	*EC* BN
11277 (12408)	**EC**	E	*EC*	BN	11291 (12535)	**EC**	E	*EC* BN
11278 (12479)	**EC**	E	*EC*	BN	11292 (12451)	**EC**	E	*EC* BN
11279 (12521)	**EC**	E	*EC*	BN	11293 (12536)	**EC**	E	*EC* BN
11280 (12523)	**EC**	E	*EC*	BN	11294 (12529)	**AL**	E	*EC* BN
11281 (12418)	**EC**	E	*EC*	BN	11295 (12475)	**EC**	E	*EC* BN
11282 (12524)	**EC**	E	*EC*	BN	11298 (12416)	**EC**	E	*EC* BN
11283 (12435)	**EC**	E	*EC*	BN	11299 (12532)	**EC**	E	*EC* BN

OPEN FIRST (DISABLED)

Mark 4. Air conditioned. Rebuilt from Open First by Bombardier Wakefield 2003–05. 42/– 1W 1TD. BT41 bogies. ETH 6X.

Advertising livery: 11325 Sky 1 (blue).

Lot No. 31046 Metro-Cammell 1989–92. 40.7 t.

11301 (11215)	**EC**	E	*EC*	BN	11316 (11227)	**EC**	E	*EC* BN
11302 (11203)	**EC**	E	*EC*	BN	11317 (11223)	**EC**	E	*EC* BN
11303 (11211)	**EC**	E	*EC*	BN	11318 (11251)	**EC**	E	*EC* BN
11304 (11257)	**EC**	E	*EC*	BN	11319 (11247)	**EC**	E	*EC* BN
11305 (11261)	**EC**	E	*EC*	BN	11320 (11255)	**EC**	E	*EC* BN
11306 (11276)	**EC**	E	*EC*	BN	11321 (11245)	**EC**	E	*EC* BN
11307 (11217)	**EC**	E	*EC*	BN	11322 (11228)	**EC**	E	*EC* BN
11308 (11263)	**EC**	E	*EC*	BN	11323 (11235)	**EC**	E	*EC* BN
11309 (11259)	**EC**	E	*EC*	BN	11324 (11253)	**EC**	E	*EC* BN
11310 (11272)	**EC**	E	*EC*	BN	11325 (11231)	**AL**	E	*EC* BN
11311 (11221)	**EC**	E	*EC*	BN	11326 (11206)	**EC**	E	*EC* BN
11312 (11225)	**EC**	E	*EC*	BN	11327 (11236)	**EC**	E	*EC* BN
11313 (11210)	**EC**	E	*EC*	BN	11328 (11274)	**EC**	E	*EC* BN
11314 (11207)	**EC**	E	*EC*	BN	11329 (11243)	**EC**	E	*EC* BN
11315 (11238)	**EC**	E	*EC*	BN	11330 (11249)	**EC**	E	*EC* BN

OPEN FIRST

Mark 4. Air conditioned. Rebuilt from Open First by Bombardier Wakefield 2003–05. Separate area for 7 smokers, although smoking is no longer allowed. 46/– 1W 1TD. BT41 bogies. ETH 6X.

Advertising livery: 11425 Sky 1 (blue).

Lot No. 31046 Metro-Cammell 1989–92. 42.1 t.

11401	(11214)	**EC**	E	*EC*	BN	11416	(11254)	**EC**	E	*EC*	BN
11402	(11216)	**EC**	E	*EC*	BN	11417	(11226)	**EC**	E	*EC*	BN
11403	(11258)	**EC**	E	*EC*	BN	11418	(11222)	**EC**	E	*EC*	BN
11404	(11202)	**EC**	E	*EC*	BN	11419	(11250)	**EC**	E	*EC*	BN
11405	(11204)	**EC**	E	*EC*	BN	11420	(11242)	**EC**	E	*EC*	BN
11406	(11205)	**EC**	E	*EC*	BN	11421	(11220)	**EC**	E	*EC*	BN
11407	(11256)	**EC**	E	*EC*	BN	11422	(11232)	**EC**	E	*EC*	BN
11408	(11218)	**EC**	E	*EC*	BN	11423	(11230)	**EC**	E	*EC*	BN
11409	(11262)	**EC**	E	*EC*	BN	11424	(11239)	**EC**	E	*EC*	BN
11410	(11260)	**EC**	E	*EC*	BN	11425	(11234)	**AL**	E	*EC*	BN
11411	(11240)	**EC**	E	*EC*	BN	11426	(11252)	**EC**	E	*EC*	BN
11412	(11209)	**EC**	E	*EC*	BN	11427	(11200)	**EC**	E	*EC*	BN
11413	(11212)	**EC**	E	*EC*	BN	11428	(11233)	**EC**	E	*EC*	BN
11414	(11246)	**EC**	E	*EC*	BN	11429	(11275)	**EC**	E	*EC*	BN
11415	(11208)	**EC**	E	*EC*	BN	11430	(11248)	**EC**	E	*EC*	BN

OPEN FIRST

Mark 4. Air conditioned. Converted from Kitchen Buffet Standard with new interior by Bombardier Wakefield 2005. 46/– 1T. BT41 bogies. ETH 6X.

Lot No. 31046 Metro-Cammell 1989–92. 41.3 t.

11998	(10314)	**EC**	E	*EC*	BN		11999	(10316)	**EC**	E	*EC*	BN

OPEN STANDARD

Mark 3A. Air conditioned. All refurbished with modified seat backs and new layout and further refurbished with new seat trim. –/76 2T (s –/70 2T 1W, t –/72 2T, z –/70 1TD 1T 2W. BT10 bogies. d. ETH 6X.

h   Greater Anglia modified coaches with 8 Compin Pegasus seats at saloon ends for "priority" use and more unidirectional seating. –/80 2T.

§   First Great Western Sleeper "day coaches" that have been fitted with former HST First Class seats to a 2+1 layout and are effectively unclassified. –/45(2) 2T 1W.

12170/171 were converted from Open Composites 11909/910, formerly Open Firsts 11009/010.

Non-standard livery: 12142 GWR brown (trial livery).

12005–167. Lot No. 30877 Derby 1975–77. 34.3 t.
12170/171. Lot No. 30878 Derby 1975–76. 34.3 t.

12005 h	**1**	P	*GA*	NC		12087 s	**V**	DR		BH
12008	**V**	P		LM		12089	**1**	P	*GA*	NC
12009 h	**GA**	P	*GA*	NC		12090 h	**GA**	P	*GA*	NC
12011	**VT**	P	*VW*	WB		12091 h	**GA**	P	*GA*	NC
12012 h	**1**	P	*GA*	NC		12092	**1**	P		LM
12013 h	**GA**	P	*GA*	NC		12093 h	**1**	P	*GA*	NC
12015 h	**GA**	P	*GA*	NC		12094	**V**	AV	*CR*	AL
12016	**1**	P		LM		12095	**V**	P		LM
12017	**BG**	AV	*CR*	AL		12097 h	**1**	P	*GA*	NC
12019 h	**GA**	P	*GA*	NC		12098	**1**	P	*GA*	NC
12021	**GA**	P	*GA*	NC		12099 h	**1**	P	*GA*	NC
12022	**V**	P		LM		12100 §	**FD**	P	*GW*	PZ
12024 h	**1**	P	*GA*	NC		12101 s	**V**	P		LM
12026 h	**1**	P	*GA*	NC		12103	**1**	P	*GA*	NC
12027 h	**1**	P	*GA*	NC		12104	**V**	AV		LM
12029	**V**	P		LM		12105 h	**1**	P	*GA*	NC
12030 h	**1**	P	*GA*	NC		12107 h	**1**	P	*GA*	NC
12031	**1**	P	*GA*	NC		12108	**1**	P	*GA*	NC
12032 h	**1**	P	*GA*	NC		12109 h	**1**	P	*GA*	NC
12034	**1**	P	*GA*	NC		12110 h	**1**	P	*GA*	NC
12035 h	**GA**	P	*GA*	NC		12111	**1**	P	*GA*	NC
12036 s	**V**	P		LM		12114 h	**1**	P	*GA*	NC
12037 h	**1**	P	*GA*	NC		12115 h	**1**	P	*GA*	NC
12040 h	**GA**	P	*GA*	NC		12116 h	**GA**	P	*GA*	NC
12041 h	**1**	P	*GA*	NC		12118	**GA**	P	*GA*	NC
12042 h	**1**	P	*GA*	NC		12119 t	**BG**	AV	*CR*	AL
12043 s	**BG**	AV	*CR*	AL		12120 h	**1**	P	*GA*	NC
12045	**V**	P		LM		12122 z	**VT**	P	*VW*	WB
12046 h	**1**	P	*GA*	NC		12125 h	**1**	P	*GA*	NC
12047 z	**V**	DR		BH		12126 h	**1**	P	*GA*	NC
12049	**1**	P	*GA*	NC		12129 h	**GA**	P	*GA*	NC
12051 h	**GA**	P	*GA*	NC		12130 h	**1**	P	*GA*	NC
12054 s	**BG**	AV	*CR*	AL		12132	**GA**	P	*GA*	NC
12056 h	**1**	P	*GA*	NC		12133	**VT**	P	*VW*	WB
12057 h	**1**	P	*GA*	NC		12134	**V**	DR		BH
12058	**V**	AV		LM		12137 h	**1**	P	*GA*	NC
12060 h	**1**	P	*GA*	NC		12138	**VT**	P	*VW*	WB
12061 h	**1**	P	*GA*	NC		12139	**NC**	P	*GA*	NC
12062 h	**1**	P	*GA*	NC		12141	**1**	P	*GA*	NC
12063	**1**	DR		BH		12142 z	**0**	P		ZH
12064 h	**1**	P	*GA*	NC		12143	**1**	P	*GA*	NC
12065	**1**	DR		BH		12144 s	**V**	P		LM
12066 h	**1**	P	*GA*	NC		12146	**1**	P	*GA*	NC
12067	**1**	P	*GA*	NC		12147	**1**	P	*GA*	NC
12073 h	**1**	P	*GA*	NC		12148	**GA**	P	*GA*	NC
12078	**VT**	P	*VW*	WB		12150 h	**1**	P	*GA*	NC
12079	**1**	P	*GA*	NC		12151	**1**	P	*GA*	NC
12081	**1**	P	*GA*	NC		12153	**GA**	P	*GA*	NC
12082 h	**1**	P	*GA*	NC		12154 h	**GA**	P	*GA*	NC
12083	**V**	P		LM		12156	**V**	P		LM
12084 h	**1**	P	*GA*	NC		12159	**GA**	P	*GA*	NC

12160	s	**V**	P		LM	12166		**1**	P	*GA*	NC
12161	§	**FD**	P	*GW*	PZ	12167	h	**1**	P	*GA*	NC
12163		**V**	P		LM	12170		**GA**	P	*GA*	NC
12164		**1**	P	*GA*	NC	12171		**GA**	P	*GA*	NC
12165		**V**	AV		LM						

OPEN STANDARD

Mark 3A (†) or Mark 3B. Air conditioned. BT10 bogies. d. ETH 6X. Converted from Mark 3A or 3B Open Firsts. 12176–181 fitted with new Grammer seating. –/70 2T 1W. 12182–185 under conversion, details awaited.

12176–181/185. Mark 3B. Lot No. 30982 Derby 1985. 38.5 t.
12182–184. Mark 3A. Lot No. 30878 Derby 1975–76.   t .

12176	(11064)		**AW**	AV	*AW*	CF
12177	(11065)		**AW**	AV	*AW*	CF
12178	(11071)		**AW**	AV	*AW*	CF
12179	(11083)		**AW**	AV	*AW*	CF
12180	(11084)		**AW**	AV	*AW*	CF
12181	(11086)		**AW**	AV	*AW*	CF
12182	(11013)	†		AV		CP
12183	(11027)	†		AV		CP
12184	(11044)	†		AV		CP
12185	(11089)			AV		CP

OPEN STANDARD (END)

Mark 4. Air conditioned. Rebuilt with new interior by Bombardier Wakefield 2003–05. Separate area for 26 smokers, although smoking is no longer allowed. –/76 1T. BT41 bogies. ETH 6X.

12232 was converted from the original 12405.

Advertising livery: 12217 Sky 1 (blue).

12200–231. Lot No. 31047 Metro-Cammell 1989–91. 39.5 t.
12232. Lot No. 31049 Metro-Cammell 1989–92. 39.5 t.

12200	**EC**	E	*EC*	BN	12217	**AL**	E	*EC*	BN
12201	**EC**	E	*EC*	BN	12218	**EC**	E	*EC*	BN
12202	**EC**	E	*EC*	BN	12219	**EC**	E	*EC*	BN
12203	**EC**	E	*EC*	BN	12220	**EC**	E	*EC*	BN
12204	**EC**	E	*EC*	BN	12222	**EC**	E	*EC*	BN
12205	**EC**	E	*EC*	BN	12223	**EC**	E	*EC*	BN
12207	**EC**	E	*EC*	BN	12224	**EC**	E	*EC*	BN
12208	**EC**	E	*EC*	BN	12225	**EC**	E	*EC*	BN
12209	**EC**	E	*EC*	BN	12226	**EC**	E	*EC*	BN
12210	**EC**	E	*EC*	BN	12227	**EC**	E	*EC*	BN
12211	**EC**	E	*EC*	BN	12228	**EC**	E	*EC*	BN
12212	**EC**	E	*EC*	BN	12229	**EC**	E	*EC*	BN
12213	**EC**	E	*EC*	BN	12230	**EC**	E	*EC*	BN
12214	**EC**	E	*EC*	BN	12231	**EC**	E	*EC*	BN
12215	**EC**	E	*EC*	BN	12232	**EC**	E	*EC*	BN
12216	**EC**	E	*EC*	BN					

OPEN STANDARD (DISABLED)

Mark 4. Air conditioned. Rebuilt with new interior by Bombardier Wakefield 2003–05. –/68 2W 1TD. BT41 bogies. ETH 6X.

12331 was converted from Open Standard 12531.

Advertising livery: 12322 Sky 1 (blue).

12300–330. Lot No. 31048 Metro-Cammell 1989–91. 39.4 t.
12331. Lot No. 31049 Metro-Cammell 1989–92. 39.4 t.

12300	**EC**	E	*EC*	BN	12317	**EC**	E	*EC*	BN
12301	**EC**	E	*EC*	BN	12318	**EC**	E	*EC*	BN
12302	**EC**	E	*EC*	BN	12319	**EC**	E	*EC*	BN
12303	**EC**	E	*EC*	BN	12320	**EC**	E	*EC*	BN
12304	**EC**	E	*EC*	BN	12321	**EC**	E	*EC*	BN
12305	**EC**	E	*EC*	BN	12322	**AL**	E	*EC*	BN
12307	**EC**	E	*EC*	BN	12323	**EC**	E	*EC*	BN
12308	**EC**	E	*EC*	BN	12324	**EC**	E	*EC*	BN
12309	**EC**	E	*EC*	BN	12325	**EC**	E	*EC*	BN
12310	**EC**	E	*EC*	BN	12326	**EC**	E	*EC*	BN
12311	**EC**	E	*EC*	BN	12327	**EC**	E	*EC*	BN
12312	**EC**	E	*EC*	BN	12328	**EC**	E	*EC*	BN
12313	**EC**	E	*EC*	BN	12329	**EC**	E	*EC*	BN
12315	**EC**	E	*EC*	BN	12330	**EC**	E	*EC*	BN
12316	**EC**	E	*EC*	BN	12331	**EC**	E	*EC*	BN

OPEN STANDARD

Mark 4. Air conditioned. Rebuilt with new interior by Bombardier Wakefield 2003–05. –/76 1T. BT41 bogies. ETH 6X.

12405 is the second coach to carry that number. It was built from the bodyshell originally intended for 12221. The original 12405 is now 12232.

Advertising livery: 12446, 12464, 12519 Sky 1 (blue).

Lot No. 31049 Metro-Cammell 1989–92. 40.8 t.

12400	**EC**	E	*EC*	BN	12420	**EC**	E	*EC*	BN
12401	**EC**	E	*EC*	BN	12421	**EC**	E	*EC*	BN
12402	**EC**	E	*EC*	BN	12422	**EC**	E	*EC*	BN
12403	**EC**	E	*EC*	BN	12423	**EC**	E	*EC*	BN
12404	**EC**	E	*EC*	BN	12424	**EC**	E	*EC*	BN
12405	**EC**	E	*EC*	BN	12425	**EC**	E	*EC*	BN
12406	**EC**	E	*EC*	BN	12426	**EC**	E	*EC*	BN
12407	**EC**	E	*EC*	BN	12427	**EC**	E	*EC*	BN
12409	**EC**	E	*EC*	BN	12428	**EC**	E	*EC*	BN
12410	**EC**	E	*EC*	BN	12429	**EC**	E	*EC*	BN
12411	**EC**	E	*EC*	BN	12430	**EC**	E	*EC*	BN
12414	**EC**	E	*EC*	BN	12431	**EC**	E	*EC*	BN
12415	**EC**	E	*EC*	BN	12432	**EC**	E	*EC*	BN
12417	**EC**	E	*EC*	BN	12433	**EC**	E	*EC*	BN
12419	**EC**	E	*EC*	BN	12434	**EC**	E	*EC*	BN

12436	**EC**	E	*EC*	BN	12467	**EC**	E	*EC*	BN
12437	**EC**	E	*EC*	BN	12468	**EC**	E	*EC*	BN
12438	**EC**	E	*EC*	BN	12469	**EC**	E	*EC*	BN
12439	**EC**	E	*EC*	BN	12470	**EC**	E	*EC*	BN
12440	**EC**	E	*EC*	BN	12471	**EC**	E	*EC*	BN
12441	**EC**	E	*EC*	BN	12472	**EC**	E	*EC*	BN
12442	**EC**	E	*EC*	BN	12473	**EC**	E	*EC*	BN
12443	**EC**	E	*EC*	BN	12474	**EC**	E	*EC*	BN
12444	**EC**	E	*EC*	BN	12476	**EC**	E	*EC*	BN
12445	**EC**	E	*EC*	BN	12477	**EC**	E	*EC*	BN
12446	**AL**	E	*EC*	BN	12478	**EC**	E	*EC*	BN
12447	**EC**	E	*EC*	BN	12480	**EC**	E	*EC*	BN
12448	**EC**	E	*EC*	BN	12481	**EC**	E	*EC*	BN
12449	**EC**	E	*EC*	BN	12483	**EC**	E	*EC*	BN
12450	**EC**	E	*EC*	BN	12484	**EC**	E	*EC*	BN
12452	**EC**	E	*EC*	BN	12485	**EC**	E	*EC*	BN
12453	**EC**	E	*EC*	BN	12486	**EC**	E	*EC*	BN
12454	**EC**	E	*EC*	BN	12488	**EC**	E	*EC*	BN
12455	**EC**	E	*EC*	BN	12489	**EC**	E	*EC*	BN
12456	**EC**	E	*EC*	BN	12513	**EC**	E	*EC*	BN
12457	**EC**	E	*EC*	BN	12514	**EC**	E	*EC*	BN
12458	**EC**	E	*EC*	BN	12515	**EC**	E	*EC*	BN
12459	**EC**	E	*EC*	BN	12518	**EC**	E	*EC*	BN
12460	**EC**	E	*EC*	BN	12519	**AL**	E	*EC*	BN
12461	**EC**	E	*EC*	BN	12520	**EC**	E	*EC*	BN
12462	**EC**	E	*EC*	BN	12522	**EC**	E	*EC*	BN
12463	**EC**	E	*EC*	BN	12526	**EC**	E	*EC*	BN
12464	**AL**	E	*EC*	BN	12533	**EC**	E	*EC*	BN
12465	**EC**	E	*EC*	BN	12534	**EC**	E	*EC*	BN
12466	**EC**	E	*EC*	BN	12538	**EC**	E	*EC*	BN

OPEN STANDARD

Mark 3A. Air conditioned. Rebuilt for Chiltern Railways 2011–13 and fitted with sliding plug doors and toilets with retention tanks. Original InterCity 70 seating retained but mainly arranged around tables. –/72(6) or * –/69(4) 1T. BT10 bogies. ETH 6X.

12602–609/614–616/618/620. Lot No. 30877 Derby 1975–77. 36.2 t (* 37.1 t).
12601/613/617–619/621/623/625/627. Lot No. 30878 Derby 1975–76. 36.2 t (* 37.1 t).

12602	(12072)		**CM**	AV	*CR*	AL
12603	(12053)	*	**CM**	AV	*CR*	AL
12604	(12131)		**CM**	AV	*CR*	AL
12605	(11040)	*	**CM**	AV	*CR*	AL
12606	(12048)		**CM**	AV	*CR*	AL
12607	(12038)	*	**CM**	AV	*CR*	AL
12608	(12069)		**CM**	AV	*CR*	AL
12609	(12014)	*	**CM**	AV	*CR*	AL
12610	(12117)		**CM**	AV	*CR*	AL
12613	(11042, 12173)	*	**CM**	AV	*CR*	AL

12614 (12145)		**CM**	AV	*CR*	AL	
12615 (12059)	*	**CM**	AV	*CR*	AL	
12616 (12127)		**CM**	AV	*CR*	AL	
12617 (11052, 12174)	*	**CM**	AV	*CR*	AL	
12618 (11008, 12169)		**CM**	AV	*CR*	AL	
12619 (11058, 12175)	*	**CM**	AV	*CR*	AL	
12620 (12124)		**CM**	AV	*CR*	AL	
12621 (11046)	*	**CM**	AV	*CR*	AL	
12623 (11019)	*	**CM**	AV	*CR*	AL	
12625 (11030)	*	**CM**	AV	*CR*	AL	
12627 (11054)	*	**CM**	AV	*CR*	AL	

CORRIDOR FIRST

Mark 1. 42/– 2T. B4 bogies. ETH 3.

Lot No. 30381 Swindon 1959. 33 t.

13227	x	**CH**	LS		CL		13230	xk	**M**	BK	*BK*	BT
13229	xk	**M**	BK	*BK*	BT							

OPEN FIRST

Mark 1 converted from Corridor First in 2013–14. 42/– 2T. Commonwealth bogies. ETH 3.

Lot No. 30667 Swindon 1962. 35 t.

13320	x	**M**	WC	*WC*	CS	ANNA

CORRIDOR FIRST

Mark 2A. Pressure ventilated. 42/– 2T. B4 bogies. ETH 4.

Lot No. 30774 Derby 1968. 33 t.

13440	v	**M**	WC	*WC*	CS

CORRIDOR BRAKE FIRST

Mark 1. 24/– 1T. Commonwealth bogies. ETH 2.

Lot No. 30668 Swindon 1961. 36 t.

Originally numbered in 14xxx series and then renumbered in 17xxx series.

17013	x	**PC**	LS		CM		17018	v	**CH**	VT	*VT*	TM

Name: 17013   BOTAURUS

CORRIDOR BRAKE FIRST

Mark 2A. Pressure ventilated. 24/– 1T. B4 bogies. ETH 4.

Originally numbered 14056–102. 17080/090 were numbered 35516/503 for a time when declassified.

17056/077. Lot No. 30775 Derby 1967–8. 32 t.
17080–102. Lot No. 30786 Derby 1968. 32 t.

17056	**M**	RV		CD	17090 v	**CH**	VT	TM
17077	**RV**	RV		BQ	17102	**M**	WC *WC*	CS
17080	**PC**	RA	*ST*	CS				

COUCHETTE/GENERATOR COACH

Mark 2B. Formerly part of Royal Train. Converted from Corridor Brake First built 1969. Consists of luggage accommodation, guard's compartment, 350 kW diesel generator and staff sleeping accommodation. Pressure ventilated. B5 bogies. ETH 5X.

Lot No. 30888 Wolverton 1977. 46 t.

17105 (14105, 2905)	**RB**	RV	*RV*	EH

CORRIDOR BRAKE FIRST

Mark 2D. Air conditioned. Stones equipment. 24/– 1T. B4 Bogies. ETH 5.

Lot No. 30823 Derby 1971–72. 33.5 t.

17159 (14159)	d	**DS**	DR	*DR*	KM	
17167 (14167)		**VN**	BE	*NB*	CP	MOW COP

OPEN BRAKE UNCLASSIFIED

Mark 3B. Air conditioned. Fitted with hydraulic handbrake. Used by First Great Western as Sleeper "day coaches" that have been fitted with former HST First Class seats to a 2+1 layout and are effectively unclassified. 36/– 1T. BT10 bogies. pg. d. ETH 5X.

Lot No. 30990 Derby 1986. 35.81 t.

17173	**FD**	P	*GW*	PZ	17175	**FD**	P	*GW*	PZ
17174	**FD**	P	*GW*	PZ					

CORRIDOR STANDARD

Mark 1. Each vehicle has eight compartments. Metal window frames and melamine interior panelling. Commonwealth bogies. –/48 2T. ETH 4.

Lot No. 30685 Derby 1961–62. 36 t.

Originally numbered 25756.

18756 (25756)	x		**M**	WC	*WC*	CS

CORRIDOR BRAKE COMPOSITE

Mark 1. There are two variants depending upon whether the Standard Class compartments have armrests. Each vehicle has two First Class and three Standard Class compartments. 12/18 2T (* 12/24 2T). Commonwealth bogies. ETH 2.

21241–245. Lot No. 30669 Swindon 1961–62. 36 t.
21256. Lot No. 30731 Derby 1963. 37 t.
21266–272. Lot No. 30732 Derby 1964. 37 t.

21241	x	**M**	BK	*BK*	BT	21266	x*	**M**	WC	*WC*	CS
21245	x	**M**	RV	*RV*	EH	21269	*	**CC**	RV	*RV*	EH
21256	x	**M**	WC	*WC*	CS	21272	x*	**CH**	RV	*RV*	EH

CORRIDOR BRAKE STANDARD

Mark 1. Four compartments. Lot No. 30721 has metal window frames and melamine interior panelling. –/24 1T. ETH2.

35185. Lot No. 30427 Wolverton 1959. B4 bogies. 33 t.
35459. Lot No. 30721 Wolverton 1963. Commonwealth bogies. 37 t.

| 35185 | x | **M** | BK | *BK* | BT |
| 35459 | x | **M** | WC | *WC* | CS |

CORRIDOR BRAKE GENERATOR STANDARD

Mark 1. Four compartments. Metal window frames and melamine interior panelling. Fitted with an ets generator in the former luggage compartment.

Lot No. 30721 Wolverton 1963. Commonwealth bogies. 37 t.

| 35469 | x | **CC** | RV | *RV* | EH |

BRAKE/POWER KITCHEN

Mark 2C. Pressure ventilated. Converted from Corridor Brake First (declassified to Corridor Brake Standard) built 1970. Converted by West Coast Railway Company 2000–01. Consists of 60 kVA generator, guard's compartment and electric kitchen. B5 bogies. ETH 4.

Non-standard livery: British Racing Green with gold lining.

Lot No. 30796 Derby 1969–70. 32.5 t.

| 35511 | (14130, 17130) | **0** | RA | BO |

KITCHEN CAR

Mark 1. Converted 1989/2006 from Kitchen Buffet Unclassified. Buffet and seating area replaced with additional kitchen and food preparation area. Fluorescent lighting. Commonwealth bogies. ETH 2X.

Lot No. 30628 Pressed Steel 1960–61. 39 t.

80041 (1690)	x	**M**	RV		EH
80042 (1646)		**BG**	RV	*RV*	EH

DRIVING BRAKE VAN (110 mph)

Mark 3B. Air conditioned. T4 bogies. dg. ETH 5X. Driving Brake Vans converted for use by Network Rail can be found in the Service Stock section of this book.

Non-standard livery: 82146 All over silver with DB logos.

Lot No. 31042 Derby 1988. 45.18 t.

82101	**V**	DR		BH	82125	**V**	P		LM
82102	**1**	P	*GA*	NC	82126	**VT**	P	*VW*	WB
82103	**GA**	P	*GA*	NC	82127	**1**	P	*GA*	NC
82105	**1**	P	*GA*	NC	82132	**1**	P	*GA*	NC
82106	**V**	AV		LB	82133	**1**	P	*GA*	NC
82107	**GA**	P	*GA*	NC	82136	**GA**	P	*GA*	NC
82110	**V**	AV		LM	82137	**V**	AV		LM
82112	**1**	P	*GA*	NC	82138	**V**	AV		LM
82113	**V**	AV		LM	82139	**1**	P	*GA*	NC
82114	**1**	P	*GA*	NC	82140	**V**	P		LM
82115	**B**	NR		ZN	82141	**V**	AV		LM
82116	**V**	AV		LM	82143	**1**	P	*GA*	NC
82118	**1**	P	*GA*	NC	82146	**0**	DB	*DB*	TO
82120	**V**	AV		LM	82148	**V**	AV		LM
82121	**1**	P	*GA*	NC	82150	**V**	AV		LM
82122	**V**	AV		LM	82152	**GA**	P	*GA*	NC
82123	**V**	AV		LM					

DRIVING BRAKE VAN (140 mph)

Mark 4. Air conditioned. Swiss-built (SIG) bogies. dg. ETH 6X.

Advertising liveries: 82205 Flying Scotsman (purple).

82216 Sky 1 (blue).

Lot No. 31043 Metro-Cammell 1988. 43.5 t.

82200	**EC**	E	*EC*	BN		82216	**AL**	E	*EC*	BN
82201	**EC**	E	*EC*	BN		82217	**EC**	E	*EC*	BN
82202	**EC**	E	*EC*	BN		82218	**EC**	E	*EC*	BN
82203	**EC**	E	*EC*	BN		82219	**EC**	E	*EC*	BN
82204	**EC**	E	*EC*	BN		82220	**EC**	E	*EC*	BN
82205	**AL**	E	*EC*	BN		82222	**EC**	E	*EC*	BN
82206	**EC**	E	*EC*	BN		82223	**EC**	E	*EC*	BN
82207	**EC**	E	*EC*	BN		82224	**EC**	E	*EC*	BN
82208	**EC**	E	*EC*	BN		82225	**EC**	E	*EC*	BN
82209	**EC**	E	*EC*	BN		82226	**EC**	E	*EC*	BN
82210	**EC**	E	*EC*	BN		82227	**EC**	E	*EC*	BN
82211	**EC**	E	*EC*	BN		82228	**EC**	E	*EC*	BN
82212	**EC**	E	*EC*	BN		82229	**EC**	E	*EC*	BN
82213	**EC**	E	*EC*	BN		82230	**EC**	E	*EC*	BN
82214	**EC**	E	*EC*	BN		82231	**EC**	E	*EC*	BN
82215	**EC**	E	*EC*	BN						

DRIVING BRAKE VAN (110 mph)

Mark 3B. Air conditioned. T4 bogies. dg. ETH 6X.

82301–305 originally converted 2008 for use on Wrexham & Shropshire services. Now operated by Chiltern Railways in push-pull mode with Class 67s. 82306–308 converted for Arriva Trains Wales 2011–12. 82309 converted for Chiltern Railways 2013.

g Fitted with a diesel generator.

Lot No. 31042 Derby 1988. 45.2 t.

82301	(82117)	g	**CM**	AV	*CR*	AL
82302	(82151)	g	**CM**	AV	*CR*	AL
82303	(82135)	g	**CM**	AV	*CR*	AL
82304	(82130)	g	**CM**	AV	*CR*	AL
82305	(82134)	g	**CM**	AV	*CR*	AL
82306	(82144)		**AW**	AV	*AW*	CF
82307	(82131)		**AW**	AV	*AW*	CF
82308	(82108)		**AW**	AV	*AW*	CF
82309	(82104)	g	**CM**	AV	*CR*	AL

GANGWAYED BRAKE VAN (100 mph)

Mark 1. Short frame (57 ft). Load 10 t. Adapted 199? for use as Brake Luggage Van. Guard's compartment retained and former baggage area adapted for secure stowage of passengers' luggage. B4 bogies. 100 mph. ETH 1X.

Lot No. 30162 Pressed Steel 1956–57. 30.5 t.

92904 (80867, 99554)		**VN**	BE	*NB*	CP

GENERAL UTILITY VAN (100 mph)

Mark 1. Short frame. Load 14 t. Screw couplers. Adapted 2013/2010 for use as a water carrier with 3000 gallon capacity.

Non-standard livery: 96100 GWR Brown.

96100. Lot No. 30565 Pressed Steel 1959. 30 t. B5 bogies.
96175. Lot No. 30403 York/Glasgow 1958–60. 30 t. Commonwealth bogies.

96100 (86734, 93734)	x	**0**	VT	*VT*		TM
96175 (86628, 93628)	x	**M**	WC	*WC*		CS

KITCHEN CAR

Mark 1 converted from Corridor First in 2008 with staff accommodation. Commonwealth bogies. ETH 3.

Lot No. 30667 Swindon 1961. 35 t.

99316 (13321)	x	**M**	WC	*WC*	CS

BUFFET STANDARD

Mark 1 converted from Open Standard in 2013 by the removal of two seating bays and fitting of a buffet. –/48 2T. Commonwealth bogies. ETH 4.

Lot No. 30646 Wolverton 1961. 36 t.

99318 (4912)	x	**M**	WC	*WC*	CS

KITCHEN CAR

Mark 1 converted from Corridor Standard in 2011 with staff accommodation. Commonwealth bogies. ETH 3.

Lot No. 30685 Derby 1961–62. 34 t.

99712 (18893)	x	**M**	WC	*WC*	CS

OPEN STANDARD

Mark 1 Corridor Standard rebuilt in 1997 as Open Standard using components from 4936. –/64 2T. Commonwealth bogies. ETH 4.

Lot No. 30685 Derby 1961–62. 36 t.

99722 (25806, 18806)	x	**M**	WC	*WC*	CS

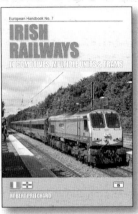

NYMR REGISTERED COACHES

These coaches are permitted to operate on the national railway network but may only be used to convey fare-paying passengers between Middlesbrough and Whitby on the Esk Valley branch as an extension of North Yorkshire Moors Railway services between Pickering and Grosmont. Only NYMR coaches currently registered for use on the national network are listed.

RESTAURANT FIRST

Mark 1. 24/–. Commonwealth bogies. ETH 2. Lot No. 30633 Swindon 1961. 42.5 t.

324	x	**PC**	NY	*NY*		NY		JOS de CRAU

BUFFET STANDARD

Mark 1. –/44 2T. Commonwealth bogies. ETH 3.

Lot No. 30520 Wolverton 1960. 38 t.

1823	v	**M**	NY	*NY*	NY

OPEN STANDARD

Mark 1. –/64 2T (* –/60 2W 2T, † –/60 3W 1T). BR Mark 1 bogies. ETH 4.

3798/3801. Lot No. 30079 York 1953. 33 t.
3860/72. Lot No. 30080 York 1954. 33 t.
3948. Lot No. 30086 Eastleigh 1954–55. 33 t.
4198/4252. Lot No. 30172 York 1956. 33 t.
4286/90. Lot No. 30207 BRCW 1956. 33 t.
4455. Lot No. 30226 BRCW 1957. 33 t.

3798	v	**M**	NY	*NY*	NY	4198	v	**CC**	NY	*NY*	NY
3801	v	**CC**	NY	*NY*	NY	4252	v*	**CC**	NY	*NY*	NY
3860	v*	**M**	NY	*NY*	NY	4286	v	**CC**	NY	*NY*	NY
3872	v†	**BG**	NY	*NY*	NY	4290	v	**M**	NY	*NY*	NY
3948	v	**CC**	NY	*NY*	NY	4455	v	**CC**	NY	*NY*	NY

OPEN STANDARD

Mark 1. –/48 2T. BR Mark 1 bogies. ETH 4.

4786. Lot No. 30376 York 1957. 33 t.
4817. Lot No. 30473 BRCW 1959. 33 t.

4786	v	**CH**	NY	*NY*	NY	4817	v	**M**	NY	*NY*	NY

OPEN STANDARD

Mark 1. Later vehicles built with Commonwealth bogies. –/64 2T. ETH 4.

Lot No. 30690 Wolverton 1961–62. Aluminium window frames. 37 t.

4990	v	**M**	NY	*NY*	NY	5000	v	**M**	NY	*NY*	NY
5029	v	**CH**	NY	*NY*	NY						

OPEN BRAKE STANDARD

Mark 1. –/39 1T. BR Mark 1 bogies. ETH 4.

Lot No. 30170 Doncaster 1956. 34 t.

| 9225 | v | **M** | NY | *NY* | NY | 9267 | v | **BG** | NY | *NY* | NY |
| 9274 | v | **M** | NY | *NY* | NY | | | | | | |

CORRIDOR COMPOSITE

Mark 1. 24/18 1T. BR Mark 1 bogies. ETH 2.

15745. Lot No. 30179 Metro Cammell 1956. 36 t.
16156. Lot No. 30665 Derby 1961. 36 t.

| 15745 | v | **M** | NY | *NY* | NY | 16156 | v | **CC** | NY | *NY* | NY |

CORRIDOR BRAKE COMPOSITE

Mark 1. Two First Class and three Standard Class compartments. 12/18 2T. BR Mark 1 bogies.

Lot No. 30185 Metro Cammell 1956. 36 t.

| 21100 | v | **CC** | NY | *NY* | NY |

CORRIDOR BRAKE STANDARD

Mark 1. –/24 1T. BR Mark 1 bogies. ETH 2.

Lot No. 30233 Gloucester 1957. 35 t.

| 35089 | v | **CC** | NY | *NY* | NY |

PULLMAN BRAKE THIRD

Built 1928 by Metropolitan Carriage & Wagon Company. –/30. Gresley bogies. 37.5 t.

| 232 | v | **PC** | NY | *NY* | NY | CAR No. 79 |

PULLMAN KITCHEN FIRST

Built by Metro-Cammell 1960–61 for ECML services. 20/– 2T. Commonwealth bogies. ETH 4. 41.2 t.

| 318 | x | **PC** | NY | *NY* | NY | ROBIN |

PULLMAN PARLOUR FIRST

Built by Metro-Cammell 1960–61 for ECML services. 29/– 2T. Commonwealth bogies. 38.5 t.

| 328 | x | **PC** | NY | *NY* | NY | OPAL |

2.2. HIGH SPEED TRAIN TRAILER CARS

HSTs consist of a number of trailer cars (usually between six and nine) with a power car at each end. All trailers are classified Mark 3 and have BT10 bogies with disc brakes and central door locking. Heating is by a 415 V three-phase supply and vehicles have air conditioning. Maximum speed is 125 mph.

The trailer cars have one standard bodyshell for both First and Standard Class, thus facilitating easy conversion from one class to the other. As built all cars had facing seating around tables with Standard Class coaches having nine bays of seats per side which did not line up with the eight windows per side. This created a new unwelcome trend in British rolling stock of seats not lining up with windows.

All vehicles underwent a mid-life refurbishment in the 1980s with Standard Class seating layouts revised to incorporate unidirectional seating in addition to facing, in a somewhat higgledy-piggledy layout where seats did not line up either side of the aisle.

A further refurbishment programme was completed in November 2000, with each Train Operating Company having a different scheme as follows:

Great Western Trains (later First Great Western). Green seat covers and extra partitions between seat bays.

Great North Eastern Railway. New ceiling lighting panels and brown seat covers. First Class vehicles had table lamps and imitation walnut plastic end panels.

Virgin CrossCountry. Green seat covers. Standard Class vehicles had four seats in the centre of each carriage replaced with a luggage stack. All have now passed to other operators.

Midland Mainline. Grey seat covers, redesigned seat squabs, side carpeting and two seats in the centre of each Standard Class carriage and one in First Class carriages replaced with a luggage stack.

Since then the remaining three operators of HSTs have embarked on separate, and very different, refurbishment projects:

Midland Mainline was first to refurbish its vehicles a second time during 2003–04. This involved fitting new fluorescent and halogen ceiling lighting, although the original seats were retained in First and Standard Class, but with blue upholstery.

London St Pancras–Sheffield/Leeds and Nottingham services are now operated by **East Midlands Trains** and in late 2009 this operator embarked on another, less radical, refurbishment which included retention of the original seats but with red upholstery in Standard Class and blue in First Class. This programme was completed in 2010.

First Great Western started a major rebuild of its HST sets in late 2006, with the programme completed in 2008. With an increased fleet of 54 sets (since reduced to 53) the new interiors feature new lighting and seating throughout. First Class seats have leather upholstery, and are made by Primarius UK. Standard Class seats are of high-back design by Grammer. A number of sets operate without a full buffet or kitchen car, instead using one of 19 TS vehicles converted to include a "mini buffet" counter for use on shorter distance services. During 2012 15 402xx or 407xx buffet vehicles were converted to Trailer Standards to make the rakes formed up to 8-cars. During 2014–15 further changes to the FGW sets include the conversion of one First Class coach from each set to Standard Class.

GNER modernised its buffet cars with new corner bars in 2004 and at the same time each HST set was made up to 9-cars with an extra Standard Class vehicle added with a disabled person's toilet.

At the end of 2006 **GNER** embarked on a major rebuild of its sets, with the work being carried out at Wabtec, Doncaster. All vehicles will have similar interiors to the Mark 4 "Mallard" fleet, with new Primarius seats throughout. The refurbishment of the 13 sets was completed by **National Express East Coast** in late 2009, these trains are now operated by **East Coast**.

Ten sets ex-Virgin CrossCountry, and some spare vehicles, were temporarily allocated to Midland Mainline for the interim service to Manchester during 2003–04 and had a facelift. Buffet cars were converted from TSB to TFKB and renumbered in the 408xx series. Most of these vehicles are now in use with First Great Western.

Open access operator **Grand Central** started operation in December 2007 with a new service from Sunderland to London King's Cross. This operator has three sets mostly using stock converted from loco-hauled Mark 3s. The seats in Standard Class have First Class spacing and in most vehicles are all facing. Increases in the number of passengers on its services has seen rakes lengthened from 5-cars to 6-cars and they can also run as 7-cars at busy times.

CrossCountry reintroduced HSTs to the Cross-Country network from 2008. Five sets were refurbished at Wabtec, Doncaster principally for use on the Plymouth–Edinburgh route. Three of these sets use stock mostly converted from loco-hauled Mark 3s and two are sets ex-Midland Mainline. The interiors are similar to refurbished East Coast sets, although the seating layout is different and one toilet per coach has been removed in favour of a luggage stack.

Operator Codes

Operator codes are shown in the heading before each set of vehicles. The first letter is always "T" for HST coaches, denoting a Trailer vehicle. The second letter denotes the passenger accommodation in that vehicle, for example "F" for First. "GS" denotes Guards accommodation and Standard Class seating. This is followed by catering provision, with "B" for buffet, and "K" for a kitchen and buffet:

TCK	Trailer Composite Kitchen		TFKB	Trailer Kitchen Buffet First
TF	Trailer First		TSB	Trailer Buffet Standard
TFB	Trailer Buffet First		TS	Trailer Standard
TGS	Trailer Guard's Standard			

TRAILER STANDARD BUFFET   TSB

19 vehicles converted at Laira 2009–10 from HST TSs for First Great Western. Refurbished with Grammer seating.

40101–119. For Lot No. details see TS. –/70 1T. 35.5 t.

40101	(42170)	**FD**	P	GW	LA
40102	(42223)	**FD**	P	GW	LA
40103	(42316)	**FD**	P	GW	LA
40104	(42254)	**FD**	P	GW	LA
40105	(42084)	**FD**	P	GW	LA
40106	(42162)	**FD**	P	GW	LA
40107	(42334)	**FD**	P	GW	LA
40108	(42314)	**FD**	P	GW	LA
40109	(42262)	**FD**	P	GW	LA
40110	(42187)	**FD**	P	GW	LA
40111	(42248)	**FD**	P	GW	LA
40112	(42336)	**FD**	P	GW	LA
40113	(42309)	**FD**	P	GW	LA
40114	(42086)	**FD**	P	GW	LA
40115	(42320)	**FD**	P	GW	LA
40116	(42147)	**FD**	P	GW	LA
40117	(42249)	**FD**	P	GW	LA
40118	(42338)	**FD**	P	GW	LA
40119	(42090)	**FD**	P	GW	LA

TRAILER BUFFET FIRST   TFB

Converted from TSB by fitting First Class seats. Renumbered from 404xx series by subtracting 200. All refurbished by First Great Western and fitted with Primarius leather seating. 23/–.

40204–221. Lot No. 30883 Derby 1976–77. 36.12 t.
40231. Lot No. 30899 Derby 1978–79. 36.12 t.

40204	**FD**	A	GW	LA		40210	**FD**	A	GW	LA
40205	**FD**	A	GW	LA		40221	**FD**	A	GW	LA
40207	**FD**	A	GW	LA		40231	**FD**	A	GW	LA

TRAILER BUFFET STANDARD   TSB

Renumbered from 400xx series by adding 400. –/33 1W.

40433 was numbered 40233 for a time when fitted with 23 First Class seats.

40402–426. Lot No. 30883 Derby 1976–77. 36.12 t.
40433. Lot No. 30899 Derby 1978–79. 36.12 t.

40402	**V**	AV		LM		40426	**GC**	A	GC	HT
40419	**V**	AV		LM		40433	**GC**	A	GC	HT
40424	**GC**	A	GC	HT					

TRAILER KITCHEN BUFFET FIRST   TFKB

These vehicles have larger kitchens than the 402xx and 404xx series vehicles, and are used in trains where a full meal service is required. They were renumbered from the 403xx series (in which the seats were unclassified) by adding 400 to the previous number. 17/–.

*   Refurbished First Great Western vehicles. Primarius leather seating.
m   Refurbished East Coast vehicles with Primarius seating.

40700–721. Lot No. 30921 Derby 1978–79. 38.16 t.
40722–735. Lot No. 30940 Derby 1979–80. 38.16 t.
40737–753. Lot No. 30948 Derby 1980–81. 38.16 t.
40754–757. Lot No. 30966 Derby 1982. 38.16 t.

40700		**ST**	P	*EM*	NL	40732		**EC**	P	*EC*	EC
40701	m	**NX**	P	*EC*	EC	40733	*	**FD**	A	*GW*	LA
40702	m	**NX**	P	*EC*	EC	40734	*	**FD**	A	*GW*	LA
40703	*	**FD**	A	*GW*	LA	40735	m	**NX**	A	*EC*	EC
40704	m	**NX**	A	*EC*	EC	40737	m	**NX**	A	*EC*	EC
40705	m	**NX**	A	*EC*	EC	40739	*	**FD**	A	*GW*	LA
40706	m	**NX**	A	*EC*	EC	40740	m	**NX**	A	*EC*	EC
40707	*	**FD**	A	*GW*	LA	40741		**ST**	P	*EM*	NL
40708	m	**NX**	P	*EC*	EC	40742	m	**NX**	A	*EC*	EC
40710	*	**FD**	A	*GW*	LA	40743	*	**FD**	A	*GW*	LA
40711	m	**NX**	A	*EC*	EC	40746		**ST**	P	*EM*	NL
40713	*	**FD**	A	*GW*	LA	40748	m	**NX**	A	*EC*	EC
40715	*	**FD**	A	*GW*	LA	40749		**ST**	P	*EM*	NL
40716	*	**FD**	A	*GW*	LA	40750		**NX**	A	*EC*	EC
40718	*	**FD**	A	*GW*	LA	40751		**ST**	P	*EM*	NL
40720	m	**NX**	A	*EC*	EC	40752	*	**FD**	A	*GW*	LA
40721	*	**FD**	A	*GW*	LA	40753		**ST**	P	*EM*	NL
40722	*	**FD**	A	*GW*	LA	40754		**ST**	P	*EM*	NL
40727	*	**FD**	A	*GW*	LA	40755	*	**FD**	A	*GW*	LA
40728		**ST**	P	*EM*	NL	40756		**ST**	P	*EM*	NL
40730		**ST**	P	*EM*	NL	40757	*	**FD**	A	*GW*	LA

TRAILER KITCHEN BUFFET FIRST   TFKB

These vehicles have been converted from TSBs in the 404xx series to be similar to the 407xx series vehicles. 17/–.

40802/804/811 were numbered 40212/232/211 for a time when fitted with 23 First Class seats.

*   Refurbished First Great Western vehicles. Primarius leather seating.
m   Refurbished East Coast vehicle with Primarius seating.

40801–803/805/808/809/811. Lot No. 30883 Derby 1976–77. 38.16 t.
40804/806/807/810. Lot No. 30899 Derby 1978–79. 38.16 t.

40801	(40027, 40427)	*	**FD**	P	*GW*	LA
40802	(40012, 40412)	*	**FD**	P	*GW*	LA

40803	(40018, 40418)	*	**FD**	P	*GW*	LA
40804	(40032, 40432)	*	**FD**	P	*GW*	LA
40805	(40020, 40420)	m	**NX**	P		EC
40806	(40029, 40429)	*	**FD**	P	*GW*	LA
40807	(40035, 40435)	*	**FD**	P	*GW*	LA
40808	(40015, 40415)	*	**FD**	P	*GW*	LA
40809	(40014, 40414)	*	**FD**	P	*GW*	LA
40810	(40030, 40430)	*	**FD**	P	*GW*	LA
40811	(40011, 40411)	*	**FD**	P	*GW*	LA

TRAILER BUFFET FIRST   TFB

Vehicles owned by First Group. Converted from TSB by First Great Western. Refurbished with Primarius leather seating. 23/–.

40900/902/904. Lot No. 30883 Derby 1976–77. 36.12 t.
40901/903. Lot No. 30899 Derby 1978–79. 36.12 t.

40900	(40022, 40422)	**FD**	FG	*GW*	LA
40901	(40036, 40436)	**FD**	FG	*GW*	LA
40902	(40023, 40423)	**FD**	FG	*GW*	LA
40903	(40037, 40437)	**FD**	FG	*GW*	LA
40904	(40001, 40401)	**FD**	FG	*GW*	LA

TRAILER FIRST   TF

As built and m 48/– 2T (m† 48/– 1T – one toilet removed for trolley space).
* Refurbished First Great Western vehicles. Primarius leather seating.
c Refurbished CrossCountry vehicles with Primarius seating and 2 tip-up seats. One toilet removed. 40/– 1TD 1W.
m Refurbished East Coast vehicles with Primarius seating.
s Fitted with centre luggage stack, disabled toilet and wheelchair space. 46/– 1T 1TD 1W.
w Wheelchair space. 47/– 2T 1W. (41154 47/– 1TD 1T 1W).
x Toilet removed for trolley space (FGW). 48/– 1T.

41004–056. Lot No. 30881 Derby 1976–77. 33.66 t.
41057–120. Lot No. 30896 Derby 1977–78. 33.66 t.
41122–148. Lot No. 30938 Derby 1979–80. 33.66 t.
41149–166. Lot No. 30947 Derby 1980. 33.66 t.
41167–169. Lot No. 30963 Derby 1982. 33.66 t.
41170. Lot No. 30967 Derby 1982. Former prototype vehicle. 33.66 t.
41176. Lot No. 30897 Derby 1977. 33.66 t.
41179/180. Lot No. 30884 Derby 1976–77. 33.66 t.
41181–184/189. Lot No. 30939 Derby 1979–80. 33.66 t.
41185–187/191. Lot No. 30969 Derby 1982. 33.66 t.
41190. Lot No. 30882 Derby 1976–77. 33.60 t.
41192. Lot No. 30897 Derby 1977–79. 33.60 t.
41193–195/41201–206. Lot No. 30878 Derby 1975–76. 34.3 t. Converted from Mark 3A Open First.

41004	*x	**FD**	A	*GW*	LA	41083	mw	**NX**	P	*EC*	EC
41005	*x	**FD**	A	*GW*	LA	41084	s	**ST**	P	*EM*	NL
41006	*w	**FD**	A	*GW*	LA	41085	*x	**FD**	FG	*GW*	LA
41008	*x	**FD**	A	*GW*	LA	41087	mt	**NX**	A	*EC*	EC
41009	*x	**FD**	A	*GW*	LA	41088	mw	**NX**	A	*EC*	EC
41010	*x	**FD**	A	*GW*	LA	41089	*w	**FD**	A	*GW*	LA
41011	*x	**FD**	A	*GW*	LA	41090	mt	**NX**	A	*EC*	EC
41012	*w	**FD**	A	*GW*	LA	41091	mt	**NX**	A	*EC*	EC
41015	*x	**FD**	A	*GW*	LA	41092	mw	**NX**	A	*EC*	EC
41016	*w	**FD**	A	*GW*	LA	41094	*w	**FD**	A	*GW*	LA
41018	*w	**FD**	A	*GW*	LA	41095	mw	**NX**	P	*EC*	EC
41020	*w	**FD**	A	*GW*	LA	41096	*x	**FD**	P	*GW*	LA
41022	*w	**FD**	A	*GW*	LA	41097	mt	**NX**	A	*EC*	EC
41024	*w	**FD**	A	*GW*	LA	41098	mw	**NX**	A	*EC*	EC
41026	c	**XC**	A	*XC*	EC	41099	mt	**NX**	A	*EC*	EC
41027	*w	**FD**	A	*GW*	LA	41100	mw	**NX**	A	*EC*	EC
41028	*w	**FD**	A	*GW*	LA	41102	*w	**FD**	A	*GW*	LA
41029	*x	**FD**	A	*GW*	LA	41103	*x	**FD**	A	*GW*	LA
41030	*w	**FD**	A	*GW*	LA	41104	*w	**FD**	A	*GW*	LA
41032	*w	**FD**	A	*GW*	LA	41106	*w	**FD**	A	*GW*	LA
41033	*x	**FD**	A	*GW*	LA	41108	*w	**FD**	P	*GW*	LA
41034	*w	**FD**	A	*GW*	LA	41109	*x	**FD**	P	*GW*	LA
41035	c	**XC**	A	*XC*	EC	41110	*w	**FD**	A	*GW*	LA
41038	*w	**FD**	A	*GW*	LA	41111		**ST**	P	*EM*	NL
41039	mt	**NX**	A	*EC*	EC	41112		**ST**	P	*EM*	NL
41040	mw	**NX**	A	*EC*	EC	41113	s	**ST**	P	*EM*	NL
41041	s	**ST**	P	*EM*	NL	41115	mt	**NX**	A	*EC*	EC
41044	mw	**NX**	A	*EC*	EC	41116	*x	**FD**	A	*GW*	LA
41046	s	**ST**	P	*EM*	NL	41117		**ST**	P	*EM*	NL
41052	*w	**FD**	A	*GW*	LA	41118	mw	**NX**	A	*EC*	EC
41055	*x	**FD**	A	*GW*	LA	41119	*x	**FD**	P	*GW*	LA
41056	*w	**FD**	A	*GW*	LA	41120	mt	**NX**	A	*EC*	EC
41057		**ST**	P	*EM*	NL	41122	*w	**FD**	A	*GW*	LA
41059	*w	**FD**	FG	*GW*	LA	41124	*w	**FD**	A	*GW*	LA
41061	w	**ST**	P	*EM*	NL	41125	*x	**FD**	A	*GW*	LA
41062		**EC**	P	*EC*	EC	41126	*w	**FD**	A	*GW*	LA
41063		**ST**	P	*EM*	NL	41128	*w	**FD**	A	*GW*	LA
41064	s	**ST**	P	*EM*	NL	41130	*w	**FD**	A	*GW*	LA
41065	*x	**FD**	A	*GW*	LA	41132	*w	**FD**	A	*GW*	LA
41066	mt	**NX**	A	*EC*	EC	41134	*w	**FD**	A	*GW*	LA
41067	s	**ST**	P	*EM*	NL	41135	*w	**FD**	A	*GW*	LA
41068	s	**ST**	P	*EM*	NL	41136	*w	**FD**	A	*GW*	LA
41069	s	**ST**	P	*EM*	NL	41137	*x	**FD**	A	*GW*	LA
41070	s	**ST**	P	*EM*	NL	41138	*w	**FD**	A	*GW*	LA
41071		**ST**	P	*EM*	NL	41139	*x	**FD**	A	*GW*	LA
41072	s	**ST**	P	*EM*	NL	41140	*w	**FD**	A	*GW*	LA
41075		**ST**	P	*EM*	NL	41142	*w	**FD**	A	*GW*	LA
41076	s	**ST**	P	*EM*	NL	41144	*w	**FD**	A	*GW*	LA
41077		**ST**	P	*EM*	NL	41145	*x	**FD**	A	*GW*	LA
41079		**ST**	P	*EM*	NL	41146	*w	**FD**	A	*GW*	LA
41081	*x	**FD**	P	*GW*	LA	41147	*x	**FD**	P	*GW*	LA

41148	*x	**FD**	P	*GW*	LA
41149	*w	**FD**	P	*GW*	LA
41150	mw	**NX**	A	*EC*	EC
41151	mt	**NX**	A	*EC*	EC
41152	mw	**NX**	A	*EC*	EC
41154	w	**EC**	P	*EC*	EC
41155	*x	**FD**	P	*GW*	LA
41156		**ST**	P	*EM*	NL
41157	*x	**FD**	A	*GW*	LA
41158	*w	**FD**	A	*GW*	LA
41159	mt	**NX**	P	*EC*	EC
41160	*w	**FD**	FG	*GW*	LA
41161	*w	**FD**	P	*GW*	LA
41162	*w	**FD**	FG	*GW*	LA
41164	mw	**NX**	A	*EC*	EC
41165	mw	**NX**	P	*EC*	EC
41166	*w	**FD**	FG	*GW*	LA
41167	*w	**FD**	FG	*GW*	LA
41168	*x	**FD**	P	*GW*	LA
41169	*w	**FD**	P	*GW*	LA

41170	(41001)		mt	**NX**	A	*EC*	EC
41176	(42142, 42352)	*w		**FD**	P	*GW*	LA
41179	(40505)	*x		**FD**	A	*GW*	LA
41180	(40511)	*w		**FD**	A	*GW*	LA
41181	(42282)	*x		**FD**	P	*GW*	LA
41182	(42278)	*w		**FD**	P	*GW*	LA
41183	(42274)	*w		**FD**	P	*GW*	LA
41184	(42270)	*x		**FD**	P	*GW*	LA
41185	(42313)		mt	**NX**	P	*EC*	EC
41186	(42312)	*w		**FD**	P	*GW*	LA
41187	(42311)	*w		**FD**	P	*GW*	LA
41189	(42298)	*w		**FD**	P	*GW*	LA
41190	(42088)		mt	**NX**	P	*EC*	EC
41191	(42318)	*x		**FD**	P	*GW*	LA
41192	(42246)	*w		**FD**	P	*GW*	LA

The following coaches have been converted from loco-hauled Mark 3 vehicles for CrossCountry.

41193	(11060)	c	**XC**	P	*XC*	EC
41194	(11016)	c	**XC**	P	*XC*	EC
41195	(11020)	c	**XC**	P	*XC*	EC

The following coaches have been converted from loco-hauled Mark 3 vehicles for Grand Central. 48/– 2T.

41201	(11045)		**GC**	A	*GC*	HT
41202	(11017)		**GC**	A	*GC*	HT
41203	(11038)		**GC**	A	*GC*	HT
41204	(11023)		**GC**	A	*GC*	HT
41205	(11036)		**GC**	A	*GC*	HT
41206	(11055)		**GC**	A	*GC*	HT

TRAILER STANDARD TS

42158 was numbered 41177 for a time when fitted with First Class seats.
42310 was numbered 41188 for a time when fitted with First Class seats.

Standard seating and m –/76 2T.
* Refurbished First Great Western vehicles. Grammer seating. –/80 2T (unless h – high density).
c Refurbished CrossCountry vehicles with Primarius seating. One toilet removed. –/82 1T.
§c Refurbished CrossCountry vehicles with Primarius seating and 2 tip-up seats. One toilet removed. –/66 1TD 2W. Note 42379/380 are –/71 1TD 1T 2W.
d FGW vehicles with disabled persons toilet and 5, 6 or 7 tip-up seats. –/68 1T 1TD 2W.
h "High density" FGW vehicles. –/84 2T.
k "High density" FGW refurbished vehicle with disabled persons toilet and 5, 6 or 7 tip-up seats. –/72 1T 1TD 2W.
m Refurbished East Coast vehicles with Primarius seating.
u Centre luggage stack (EMT) –/74 2T.
w Centre luggage stack and wheelchair space (EMT) –72 2T 1W.
† Disabled persons toilet (East Coast) –/62 1T 1TD 1W.

42003–089/362. Lot No. 30882 Derby 1976–77. 33.6 t.
42091–250. Lot No. 30897 Derby 1977–79. 33.6 t.
42251–305. Lot No. 30939 Derby 1979–80. 33.6 t.
42306–322. Lot No. 30969 Derby 1982. 33.6 t.
42323–341. Lot No. 30983 Derby 1984–85. 33.6 t.
42342/360. Lot No. 30949 Derby 1982. 33.47 t. Converted from TGS.
42343/345. Lot No. 30970 Derby 1982. 33.47 t. Converted from TGS.
42344/361. Lot No. 30964 Derby 1982. 33.47 t. Converted from TGS.
42346/347/350/351/379/380/551–564. Lot No. 30881 Derby 1976–77. 33.66 t. Converted from TF.
42348/349/363–365/381/565–570. Lot No. 30896 Derby 1977–78. 33.66 t. Converted from TF.
42354. Lot No. 30897 Derby 1977. Was TF from 1983 to 1992. 33.66 t.
42353/355–357. Lot No. 30967 Derby 1982. Ex-prototype vehicles. 33.66 t.
42366–378/382/383/401–409. Lot No. 30877 Derby 1975–77. 34.3 t. Converted from Mark 3A Open Standard.
42384. Lot No. 30896 Derby 1977–78. 33.66 t. Converted from TF.
42385/580–583. Lot No. 30947 Derby 1980. 33.66 t. Converted from TF.
42501/509/513/515/517. Lot No. 30948 Derby 1980–81. 34.8 t. Converted from TFKB.
42502/506/508/514/516. Lot No. 30940 Derby 1979–80. 34.8 t. Converted from TFKB.
42503/504/510/511. Lot No. 30921 Derby 1978–79. 34.8 t. Converted from TFKB.
42505/507/512/518/519. Lot No. 30883 Derby 1976–77. 34.8 t. Converted from TSB.
42520. Lot No. 30899 Derby 1978–79. 34.8 t. Converted from TSB.
42571–579. Lot No. 30938 Derby 1979–80. 33.66 t. Converted from TF.

42003	*h	FD	A	GW	LA
42004	*d	FD	A	GW	LA
42005	*h	FD	A	GW	LA
42006	*h	FD	A	GW	LA
42007	*d	FD	A	GW	LA
42008	*k	FD	A	GW	LA
42009	*h	FD	A	GW	LA
42010	*h	FD	A	GW	LA
42012	*k	FD	A	GW	LA
42013	*h	FD	A	GW	LA
42014	*h	FD	A	GW	LA
42015	*k	FD	A	GW	LA
42016	*h	FD	A	GW	LA
42019	*	FD	A	GW	LA
42021	*k	FD	A	GW	LA
42023	*h	FD	A	GW	LA
42024	*k	FD	A	GW	LA
42025	*h	FD	A	GW	LA
42026	*h	FD	A	GW	LA
42027	*h	FD	A	GW	LA
42028	*h	FD	A	GW	LA
42029	*h	FD	A	GW	LA
42030	*k	FD	A	GW	LA
42031	*h	FD	A	GW	LA
42032	*h	FD	A	GW	LA
42033	*	FD	A	GW	LA
42034	*	FD	A	GW	LA
42035	*	FD	A	GW	LA
42036	c	XC	A	XC	EC
42037	c	XC	A	XC	EC
42038	c	XC	A	XC	EC
42039	*h	FD	A	GW	LA
42040	*h	FD	A	GW	LA
42041	*h	FD	A	GW	LA
42042	*h	FD	A	GW	LA
42043	*h	FD	A	GW	LA
42044	*h	FD	A	GW	LA
42045	*	FD	A	GW	LA
42046	*	FD	A	GW	LA
42047	*	FD	A	GW	LA
42048	*h	FD	A	GW	LA
42049	*h	FD	A	GW	LA
42050	*h	FD	A	GW	LA
42051	c	XC	A	XC	EC
42052	c	XC	A	XC	EC
42053	c	XC	A	XC	EC
42054	*	FD	A	GW	LA
42055	*	FD	A	GW	LA
42056	*	FD	A	GW	LA
42057	m	NX	A	EC	EC
42058	m	NX	A	EC	EC
42059	m	NX	A	EC	EC
42060	*h	FD	A	GW	LA
42061	*h	FD	A	GW	LA
42062	*k	FD	A	GW	LA
42063	m	NX	A	EC	EC
42064	m	NX	A	EC	EC
42065	m	NX	A	EC	EC
42066	*k	FD	A	GW	LA
42067	*h	FD	A	GW	LA
42068	*h	FD	A	GW	LA
42069	*k	FD	A	GW	LA
42070	*h	FD	A	GW	LA
42071	*h	FD	A	GW	LA
42072	*	FD	A	GW	LA
42073	*h	FD	A	GW	LA
42074	*h	FD	A	GW	LA
42075	*	FD	A	GW	LA
42076	*	FD	A	GW	LA
42077	*	FD	A	GW	LA
42078	*	FD	A	GW	LA
42079	*h	FD	A	GW	LA
42080	*h	FD	A	GW	LA
42081	*k	FD	A	GW	LA
42083	*h	FD	A	GW	LA
42085	*h	FD	P	GW	LA
42087	*h	FD	P	GW	LA
42089	*h	FD	A	GW	LA
42091	mt	NX	A	EC	EC
42092	*k	FD	FG	GW	LA
42093	*h	FD	FG	GW	LA
42094	*h	FD	FG	GW	LA
42095	*	FD	FG	GW	LA
42096	*h	FD	A	GW	LA
42097	c	XC	A	XC	EC
42098	*h	FD	A	GW	LA
42099	*h	FD	A	GW	LA
42100	u	ST	P	EM	NL
42101	*h	FD	P	GW	LA
42102	*h	FD	P	GW	LA
42103	*k	FD	FG	GW	LA
42104	m	NX	A	EC	EC
42105	*k	FD	FG	GW	LA
42106	m	NX	A	EC	EC
42107	*	FD	A	GW	LA
42108	*h	FD	FG	GW	LA
42109	m	NX	P	EC	EC
42110	m	NX	P	EC	EC
42111	u	ST	P	EM	NL
42112	u	ST	P	EM	NL
42113	u	ST	P	EM	NL
42115	*h	FD	P	GW	LA

42116	mt	**NX**	A	*EC*	EC	42171	m	**NX**	A	*EC*	EC
42117	m	**NX**	P	*EC*	EC	42172	m	**NX**	A	*EC*	EC
42118	*h	**FD**	A	*GW*	LA	42173	*k	**FD**	P	*GW*	LA
42119	u	**ST**	P	*EM*	NL	42174	*k	**FD**	P	*GW*	LA
42120	u	**ST**	P	*EM*	NL	42175	*h	**FD**	FG	*GW*	LA
42121	u	**ST**	P	*EM*	NL	42176	*h	**FD**	FG	*GW*	LA
42122	m	**NX**	A	*EC*	EC	42177	*h	**FD**	FG	*GW*	LA
42123		**EC**	P	*EC*	EC	42178	*h	**FD**	P	*GW*	LA
42124	u	**ST**	P	*EM*	NL	42179	m	**NX**	A	*EC*	EC
42125		**EC**	P	*EC*	EC	42180	m	**NX**	A	*EC*	EC
42126	*h	**FD**	A	*GW*	LA	42181	m	**NX**	A	*EC*	EC
42127	mt	**NX**	A	*EC*	EC	42182	m	**NX**	A	*EC*	EC
42128	mt	**NX**	A	*EC*	EC	42183	*d	**FD**	A	*GW*	LA
42129	*	**FD**	A	*GW*	LA	42184	*	**FD**	A	*GW*	LA
42130	m	**NX**	P	*EC*	EC	42185	*	**FD**	A	*GW*	LA
42131	u	**ST**	P	*EM*	NL	42186	m	**NX**	A	*EC*	EC
42132	w	**ST**	P	*EM*	NL	42188	mt	**NX**	A	*EC*	EC
42133	u	**ST**	P	*EM*	NL	42189	mt	**NX**	A	*EC*	EC
42134	m	**NX**	A	*EC*	EC	42190	m	**NX**	A	*EC*	EC
42135	u	**ST**	P	*EM*	NL	42191	m	**NX**	A	*EC*	EC
42136	u	**ST**	P	*EM*	NL	42192	m	**NX**	A	*EC*	EC
42137	u	**ST**	P	*EM*	NL	42193	m	**NX**	A	*EC*	EC
42138	*k	**FD**	A	*GW*	LA	42194	w	**ST**	P	*EM*	NL
42139	u	**ST**	P	*EM*	NL	42195	*k	**FD**	P	*GW*	LA
42140	u	**ST**	P	*EM*	NL	42196	*k	**FD**	A	*GW*	LA
42141	u	**ST**	P	*EM*	NL	42197	*h	**FD**	A	*GW*	LA
42143	*	**FD**	A	*GW*	LA	42198	m	**NX**	A	*EC*	EC
42144	*	**FD**	A	*GW*	LA	42199	m	**NX**	A	*EC*	EC
42145	*	**FD**	A	*GW*	LA	42200	*d	**FD**	A	*GW*	LA
42146	m	**NX**	A	*EC*	EC	42201	*k	**FD**	A	*GW*	LA
42148	u	**ST**	P	*EM*	NL	42202	*k	**FD**	A	*GW*	LA
42149	u	**ST**	P	*EM*	NL	42203	*h	**FD**	A	*GW*	LA
42150	m	**NX**	A	*EC*	EC	42204	*h	**FD**	A	*GW*	LA
42151	w	**ST**	P	*EM*	NL	42205		**EC**	P	*EC*	EC
42152	u	**ST**	P	*EM*	NL	42206	*d	**FD**	A	*GW*	LA
42153	u	**ST**	P	*EM*	NL	42207	*d	**FD**	A	*GW*	LA
42154	m	**NX**	A	*EC*	EC	42208	*	**FD**	A	*GW*	LA
42155	w	**ST**	P	*EM*	NL	42209	*	**FD**	A	*GW*	LA
42156	u	**ST**	P	*EM*	NL	42210		**EC**	P	*EC*	EC
42157	u	**ST**	P	*EM*	NL	42211	*k	**FD**	A	*GW*	LA
42158	m	**NX**	A	*EC*	EC	42212	*h	**FD**	A	*GW*	LA
42159	mt	**NX**	P	*EC*	EC	42213	*h	**FD**	A	*GW*	LA
42160	m	**NX**	P	*EC*	EC	42214	*h	**FD**	A	*GW*	LA
42161	mt	**NX**	A	*EC*	EC	42215	m	**NX**	A	*EC*	EC
42163	m	**NX**	P	*EC*	EC	42216	*h	**FD**	A	*GW*	LA
42164		**ST**	P	*EM*	NL	42217	*k	**FD**	P	*GW*	LA
42165		**ST**	P	*EM*	NL	42218	*h	**FD**	P	*GW*	LA
42166	*h	**FD**	P	*GW*	LA	42219	m	**NX**	A	*EC*	EC
42167	*h	**FD**	FG	*GW*	LA	42220	w	**ST**	P	*EM*	NL
42168	*h	**FD**	FG	*GW*	LA	42221	*h	**FD**	A	*GW*	LA
42169	*h	**FD**	FG	*GW*	LA	42222	*h	**FD**	P	*GW*	LA

42224	*k	**FD**	P	*GW*	LA		42280	*	**FD**	A	*GW*	LA
42225	u	**ST**	P	*EM*	NL		42281	*	**FD**	A	*GW*	LA
42226	m	**NX**	A	*EC*	EC		42283	*h	**FD**	A	*GW*	LA
42227	u	**ST**	P	*EM*	NL		42284	*h	**FD**	A	*GW*	LA
42228	m	**NX**	A	*EC*	EC		42285	*h	**FD**	A	*GW*	LA
42229	u	**ST**	P	*EM*	NL		42286	mt	**NX**	P	*EC*	EC
42230	u	**ST**	P	*EM*	NL		42287	*k	**FD**	A	*GW*	LA
42231	*h	**FD**	FG	*GW*	LA		42288	*h	**FD**	A	*GW*	LA
42232	*h	**FD**	FG	*GW*	LA		42289	*h	**FD**	A	*GW*	LA
42233	*h	**FD**	FG	*GW*	LA		42290	c	**XC**	P	*XC*	EC
42234	c	**XC**	P	*XC*	EC		42291	*d	**FD**	A	*GW*	LA
42235	m	**NX**	A	*EC*	EC		42292	*d	**FD**	A	*GW*	LA
42236	*h	**FD**	A	*GW*	LA		42293	*	**FD**	A	*GW*	LA
42237	m	**NX**	P	*EC*	EC		42294	*	**FD**	P	*GW*	LA
42238	mt	**NX**	A	*EC*	EC		42295	*d	**FD**	A	*GW*	LA
42239	mt	**NX**	A	*EC*	EC		42296	*	**FD**	A	*GW*	LA
42240	m	**NX**	A	*EC*	EC		42297	*	**FD**	A	*GW*	LA
42241	m	**NX**	A	*EC*	EC		42299	*d	**FD**	A	*GW*	LA
42242	m	**NX**	A	*EC*	EC		42300	*	**FD**	A	*GW*	LA
42243	m	**NX**	A	*EC*	EC		42301	*	**FD**	A	*GW*	LA
42244	m	**NX**	A	*EC*	EC		42302	*k	**FD**	FG	*GW*	LA
42245	*	**FD**	A	*GW*	LA		42303	*h	**FD**	FG	*GW*	LA
42247	*h	**FD**	P	*GW*	LA		42304	*h	**FD**	FG	*GW*	LA
42250	*	**FD**	A	*GW*	LA		42305	*h	**FD**	FG	*GW*	LA
42251	*k	**FD**	A	*GW*	LA		42306	m	**NX**	P	*EC*	EC
42252	*	**FD**	A	*GW*	LA		42307	m	**NX**	P	*EC*	EC
42253	*	**FD**	A	*GW*	LA		42308	*h	**FD**	P	*GW*	LA
42255	*d	**FD**	A	*GW*	LA		42310	*k	**FD**	P	*GW*	LA
42256	*	**FD**	A	*GW*	LA		42315	*h	**FD**	P	*GW*	LA
42257	*	**FD**	A	*GW*	LA		42317	*k	**FD**	P	*GW*	LA
42258	*h	**FD**	P	*GW*	LA		42319	*h	**FD**	P	*GW*	LA
42259	*k	**FD**	A	*GW*	LA		42321	*h	**FD**	P	*GW*	LA
42260	*h	**FD**	A	*GW*	LA		42322	m	**NX**	P	*EC*	EC
42261	*h	**FD**	A	*GW*	LA		42323	m	**NX**	A	*EC*	EC
42263	*	**FD**	A	*GW*	LA		42325	*	**FD**	A	*GW*	LA
42264	*k	**FD**	A	*GW*	LA		42326	m	**NX**	P	*EC*	EC
42265	*	**FD**	A	*GW*	LA		42327	w	**ST**	P	*EM*	NL
42266	*k	**FD**	P	*GW*	LA		42328	w	**ST**	P	*EM*	NL
42267	*d	**FD**	A	*GW*	LA		42329	w	**ST**	P	*EM*	NL
42268	*d	**FD**	A	*GW*	LA		42330	m	**NX**	P	*EC*	EC
42269	*	**FD**	A	*GW*	LA		42331	u	**ST**	P	*EM*	NL
42271	*k	**FD**	A	*GW*	LA		42332	*	**FD**	A	*GW*	LA
42272	*h	**FD**	A	*GW*	LA		42333	*	**FD**	A	*GW*	LA
42273	*h	**FD**	A	*GW*	LA		42335	w	**EC**	P	*EC*	EC
42275	*d	**FD**	A	*GW*	LA		42337	w	**ST**	P	*EM*	NL
42276	*	**FD**	A	*GW*	LA		42339	w	**ST**	P	*EM*	NL
42277	*	**FD**	A	*GW*	LA		42340	m	**NX**	A	*EC*	EC
42279	*d	**FD**	A	*GW*	LA		42341	u	**ST**	P	*EM*	NL

42342	(44082)	c	**XC**	A	*XC*	EC
42343	(44095)	*	**FD**	A	*GW*	LA

42344	(44092)	*k	**FD**	A	*GW*	LA
42345	(44096)	*d	**FD**	A	*GW*	LA
42346	(41053)	*h	**FD**	A	*GW*	LA
42347	(41054)	*k	**FD**	A	*GW*	LA
42348	(41073)	*k	**FD**	A	*GW*	LA
42349	(41074)	*h	**FD**	A	*GW*	LA
42350	(41047)	*	**FD**	A	*GW*	LA
42351	(41048)	*	**FD**	A	*GW*	LA
42353	(42001, 41171)	*k	**FD**	FG	*GW*	LA
42354	(42114, 41175)	m	**NX**	A	*EC*	EC
42355	(42000, 41172)	m	**NX**	A	*EC*	EC
42356	(42002, 41173)	*k	**FD**	A	*GW*	LA
42357	(41002, 41174)	m	**NX**	A	*EC*	EC
42360	(44084, 45084)	*h	**FD**	A	*GW*	LA
42361	(44099, 42000)	*h	**FD**	A	*GW*	LA
42362	(42011, 41178)	*h	**FD**	A	*GW*	LA
42363	(41082)	m†	**NX**	A	*EC*	EC
42364	(41080)	*k	**FD**	P	*GW*	LA
42365	(41107)	*h	**FD**	P	*GW*	LA

42366–378 have been converted from loco-hauled Mark 3 vehicles for CrossCountry and 42382/383 for First Great Western.

42366	(12007)	§c	**XC**	P	*XC*	EC
42367	(12025)	c	**XC**	P	*XC*	EC
42368	(12028)	c	**XC**	P	*XC*	EC
42369	(12050)	c	**XC**	P	*XC*	EC
42370	(12086)	c	**XC**	P	*XC*	EC
42371	(12052)	§c	**XC**	P	*XC*	EC
42372	(12055)	c	**XC**	P	*XC*	EC
42373	(12071)	c	**XC**	P	*XC*	EC
42374	(12075)	c	**XC**	P	*XC*	EC
42375	(12113)	c	**XC**	P	*XC*	EC
42376	(12085)	§c	**XC**	P	*XC*	EC
42377	(12102)	c	**XC**	P	*XC*	EC
42378	(12123)	c	**XC**	P	*XC*	EC
42379	(41036)	§c	**XC**	A	*XC*	EC
42380	(41025)	§c	**XC**	A	*XC*	EC
42381	(41058)	*k	**FD**	P	*GW*	LA
42382	(12128)	*h	**FD**	P	*GW*	LA
42383	(12172)	*h	**FD**	P	*GW*	LA
42384	(41078)	w	**ST**	P	*EM*	NL
42385	(41153)	*h	**FD**	P	*GW*	LA

The following coaches have been converted from loco-hauled Mark 3 vehicles for Grand Central. They have a lower density seating layout (most seats arranged around tables). –/64 2T († –/60 1TD 1T 2W).

42401	(12149)		**GC**	A	*GC*	HT
42402	(12155)		**GC**	A	*GC*	HT
42403	(12033) †		**GC**	A	*GC*	HT
42404	(12152)		**GC**	A	*GC*	HT
42405	(12136)		**GC**	A	*GC*	HT

42406	(12112) †	**GC**	A	*GC*	HT
42407	(12044)	**GC**	A	*GC*	HT
42408	(12121)	**GC**	A	*GC*	HT
42409	(12088) †	**GC**	A	*GC*	HT

42501–520. These coaches were converted from TFKB or TSB buffet cars to TS vehicles in 2011–12 (42501–515) and 2013–14 (42516–520) at Wabtec Kilmarnock for First Great Western. –/84 1T. 34.8 t.

42501	(40344, 40744)	*	**FD**	A	*GW*	LA
42502	(40331, 40731)	*	**FD**	A	*GW*	LA
42503	(40312, 40712)	*	**FD**	A	*GW*	LA
42504	(40314, 40714)	*	**FD**	A	*GW*	LA
42505	(40428, 40228)	*	**FD**	A	*GW*	LA
42506	(40324, 40724)	*	**FD**	A	*GW*	LA
42507	(40409, 40209)	*	**FD**	A	*GW*	LA
42508	(40325, 40725)	*	**FD**	A	*GW*	LA
42509	(40336, 40736)	*	**FD**	A	*GW*	LA
42510	(40317, 40717)	*	**FD**	A	*GW*	LA
42511	(40309, 40709)	*	**FD**	A	*GW*	LA
42512	(40408, 40208)	*	**FD**	A	*GW*	LA
42513	(40338, 40738)	*	**FD**	A	*GW*	LA
42514	(40326, 40726)	*	**FD**	A	*GW*	LA
42515	(40347, 40747)	*	**FD**	A	*GW*	LA
42516	(40323, 40723)	*	**FD**	A	*GW*	LA
42517	(40345, 40745)	*	**FD**	A	*GW*	LA
42518	(40003, 40403)	*	**FD**	P	*GW*	LA
42519	(40016, 40416)	*	**FD**	P	*GW*	LA
42520	(40234, 40434)	*	**FD**	P	*GW*	LA

42551–583. These coaches were converted or are being converted from TF to TS vehicles in 2014 at Wabtec Kilmarnock for First Great Western. 42583 will be converted from TS 42385. –/80 1T. 35.5 t.

42551	(41003)	*	**FD**	A	*GW*	LA
42552	(41007)	*	**FD**	A	*GW*	LA
42553	(41009)	*	**FD**	A		
42554	(41011)	*	**FD**	A		
42555	(41015)	*	**FD**	A		
42556	(41017)	*	**FD**	A	*GW*	LA
42557	(41019)	*	**FD**	A	*GW*	LA
42558	(41021)	*	**FD**	A	*GW*	LA
42559	(41023)	*	**FD**	A	*GW*	LA
42560	(41027)	*	**FD**	A		
42561	(41031)	*	**FD**	A	*GW*	LA
42562	(41037)	*	**FD**	A	*GW*	LA
42563	(41045)	*	**FD**	FG	*GW*	LA
42564	(41051)	*	**FD**	A	*GW*	LA
42565	(41085)	*	**FD**	FG		
42566	(41086)	*	**FD**	FG	*GW*	LA
42567	(41093)	*	**FD**	A	*GW*	LA
42568	(41101)	*	**FD**	A	*GW*	LA
42569	(41105)	*	**FD**	A	*GW*	LA

42570	(41114)	*	**FD**	FG	*GW*	LA
42571	(41121)	*	**FD**	A	*GW*	LA
42572	(41123)	*	**FD**	A	*GW*	LA
42573	(41127)	*	**FD**	A	*GW*	LA
42574	(41129)	*	**FD**	A	*GW*	LA
42575	(41131)	*	**FD**	A	*GW*	LA
42576	(41133)	*	**FD**	A	*GW*	LA
42577	(41141)	*	**FD**	A	*GW*	LA
42578	(41143)	*	**FD**	A	*GW*	LA
42579	(41145)	*	**FD**	A		
42580	(41155)	*	**FD**	P		
42581	(41157)	*	**FD**	A		
42582	(41163)	*	**FD**	FG	*GW*	LA
42583	(41153, 42385)	*	**FD**	P		

TRAILER GUARD'S STANDARD   TGS

As built and m –/65 1T.
* Refurbished First Great Western vehicles. Grammer seating and toilet removed for trolley store. –/67 (unless h).
† Contains a mix of original seats and 20 new Grammer seats. –/65 1T.
c Refurbished CrossCountry vehicles with Primarius seating. –/67 1T.
h "High density" FGW vehicles. –/71.
m Refurbished East Coast vehicles with Primarius seating.
s Fitted with centre luggage stack (EMT) –/63 1T.
t Fitted with centre luggage stack –/61 1T.

44000. Lot No. 30953 Derby 1980. 33.47 t.
44001–090. Lot No. 30949 Derby 1980–82. 33.47 t.
44091–094. Lot No. 30964 Derby 1982. 33.47 t.
44097–101. Lot No. 30970 Derby 1982. 33.47 t.

44000	*h	**FD**	P	*GW*	LA		44021	c	**XC**	P	*XC*	EC
44001	*	**FD**	A	*GW*	LA		44022	*h	**FD**	A	*GW*	LA
44002	*h	**FD**	A	*GW*	LA		44023	*h	**FD**	A	*GW*	LA
44003	*h	**FD**	A	*GW*	LA		44024	*h	**FD**	A	*GW*	OO
44004	*h	**FD**	A	*GW*	LA		44025	*	**FD**	A	*GW*	LA
44005	*h	**FD**	A	*GW*	LA		44026	*h	**FD**	A	*GW*	LA
44007	*h	**FD**	A	*GW*	LA		44027	s	**ST**	P	*EM*	NL
44008	*h	**FD**	A	*GW*	LA		44028	*	**FD**	A	*GW*	LA
44009	*h	**FD**	A	*GW*	LA		44029	*	**FD**	A	*GW*	LA
44010	*h	**FD**	A	*GW*	LA		44030	*h	**FD**	A	*GW*	LA
44011	*	**FD**	A	*GW*	LA		44031	m	**NX**	A	*EC*	EC
44012	c	**XC**	A	*XC*	EC		44032	*	**FD**	A	*GW*	LA
44013	*h	**FD**	A	*GW*	LA		44033	*h	**FD**	A	*GW*	LA
44014	*h	**FD**	A	*GW*	LA		44034	*	**FD**	A	*GW*	LA
44015	*	**FD**	A	*GW*	LA		44035	*	**FD**	A	*GW*	LA
44016	*h	**FD**	A	*GW*	LA		44036	*h	**FD**	A	*GW*	LA
44017	c	**XC**	A	*XC*	EC		44037	*h	**FD**	A	*GW*	LA
44018	*	**FD**	A	*GW*	LA		44038	*	**FD**	A	*GW*	LA
44019	m	**NX**	A	*EC*	EC		44039	*	**FD**	A	*GW*	LA
44020	*h	**FD**	A	*GW*	LA		44040	*	**FD**	A	*GW*	LA

44041	s	**ST**	P	*EM*	NL
44042	*h	**FD**	P	*GW*	LA
44043	*h	**FD**	A	*GW*	LA
44044	s	**ST**	P	*EM*	NL
44045	m	**NX**	A	*EC*	EC
44046	s	**ST**	P	*EM*	NL
44047	s	**ST**	P	*EM*	NL
44048	s	**ST**	P	*EM*	NL
44049	*	**FD**	A	*GW*	LA
44050	m	**NX**	P	*EC*	EC
44051	s	**ST**	P	*EM*	NL
44052	c	**XC**	P	*XC*	EC
44054	s	**ST**	P	*EM*	NL
44055	*h	**FD**	FG	*GW*	LA
44056	m	**NX**	A	*EC*	EC
44057	m	**NX**	P	*EC*	EC
44058	m	**NX**	A	*EC*	EC
44059	*	**FD**	A	*GW*	LA
44060	*h	**FD**	P	*GW*	LA
44061	m	**NX**	A	*EC*	EC
44063	m	**NX**	A	*EC*	EC
44064	*h	**FD**	A	*GW*	LA
44065	t	**V**	AV		LM
44066	*	**FD**	A	*GW*	LA
44067	*h	**FD**	A	*GW*	LA
44068	*h	**FD**	FG	*GW*	LA
44069	*h	**FD**	P	*GW*	LA
44070	s	**ST**	P	*EM*	NL
44071	s	**ST**	P	*EM*	NL
44072	c	**XC**	P	*XC*	EC
44073	t	**EC**	P	*EC*	EC
44074	*h	**FD**	FG	*GW*	LA
44075	m	**NX**	P	*EC*	EC
44076	*h	**FD**	FG	*GW*	LA
44077	m	**NX**	A	*EC*	EC
44078	*h	**FD**	P	*GW*	LA
44079	*h	**FD**	P	*GW*	LA
44080	m	**NX**	A	*EC*	EC
44081	*h	**FD**	P	*GW*	LA
44083	*h	**FD**	P	*GW*	LA
44085	s	**ST**	P	*EM*	NL
44086	*	**FD**	A	*GW*	LA
44088	t	**V**	AV		LM
44089	t	**V**	AV		LM
44090	*h	**FD**	P	*GW*	LA
44091	*h	**FD**	P	*GW*	LA
44093	*h	**FD**	A	*GW*	LA
44094	m	**NX**	A	*EC*	EC
44097	*h	**FD**	P	*GW*	LA
44098	m	**NX**	A	*EC*	EC
44100	*h	**FD**	FG	*GW*	PM
44101	*h	**FD**	P	*GW*	LA

TRAILER COMPOSITE KITCHEN   TCK

Converted from Mark 3A Open Standard. Refurbished CrossCountry vehicles
with Primarius seating. Small kitchen for the preparation of hot food and
stowage space for two trolleys between First and Standard Class. One toilet
removed. 30/10 1T.

45001–005. Lot No. 30877 Derby 1975–77. 34.3 t.

45001	(12004)	**XC**	P	*XC*	EC
45002	(12106)	**XC**	P	*XC*	EC
45003	(12076)	**XC**	P	*XC*	EC
45004	(12077)	**XC**	P	*XC*	EC
45005	(12080)	**XC**	P	*XC*	EC

TRAILER COMPOSITE

To be converted from TF 2014–15 for First Great Western. 24/39 1T.

46001–004. Lot No. 30881 Derby 1976–77.   t. Converted from TF.
46005–009. Lot No. 30896 Derby 1977–78.   t. Converted from TF.
46010–013. Lot No. 30938 Derby 1979–80.   t. Converted from TF.
46014. Lot No. 30963 Derby 1982.   t. Converted from TF.
46015. Lot No. 30884 Derby 1976–77.   t. Converted from TF.
46016/017. Lot No. 30939 Derby 1979–80.   t. Converted from TF.
46018. Lot No. 30969 Derby 1982.   t. Converted from TF.

46001	(41005)	*	**FD**	A
46002	(41029)	*	**FD**	A
46003	(41033)	*	**FD**	A
46004	(41055)	*	**FD**	A
46005	(41065)	*	**FD**	A
46006	(41081)	*	**FD**	P
46007	(41096)	*	**FD**	P
46008	(41109)	*	**FD**	P
46009	(41119)	*	**FD**	P
46010	(41125)	*	**FD**	A
46011	(41139)	*	**FD**	A
46012	(41147)	*	**FD**	P
46013	(41148)	*	**FD**	P
46014	(41168)	*	**FD**	P
46015	(41179)	*	**FD**	A
46016	(41181)	*	**FD**	P
46017	(41184)	*	**FD**	P
46018	(41191)	*	**FD**	P

2.2. HST SET FORMATIONS

FIRST GREAT WESTERN

The largest operator of HSTs is First Great Western with 53 sets to cover 49 diagrams (of these one is a "hot spare" at Old Oak Common and one a "hot spare" at Bristol St Philip's Marsh).

The sets are split into three types, as shown below. These are 16 "low density" sets mainly used on West Country services, 19 "high density" sets with a full kitchen or buffet vehicle and 18 "super high density" sets including a TSB vehicle (with just a small corner buffet counter).

Although some sets are prefixed "OC" for Old Oak Common, for maintenance purposes all trailers are now based at Laira apart from two spare vehicles.

Set formations were in a state of flux as this book closed for press. A programme to reduce each set to just one TF instead of two was halfway through. Sets LA01–LA16/OC30–OC38/LA60–LA75 are shown as they will be formed at the conclusion of this programme, while sets OC40–OC57 (which will receive Trailer Composite vehicles numbered in the 46001–018 series in 2015) are still shown with two TF vehicles.

Number of sets: 53.
Maximum number of daily diagrams: 49.
Formations: 8-cars.
Allocation: Laira (Plymouth).
Other maintenance and servicing depots: Landore (Swansea), St Philip's Marsh (Bristol), Long Rock (Penzance).
Operation: London Paddington–Exeter/Paignton/Plymouth/(Newquay in the summer)/Penzance, Bristol, Cardiff/Swansea/West Wales, Oxford/Hereford/Great Malvern/Cheltenham Spa.

Set	L	K	F	E	D	C	B	A	density
LA01	41024	40755	42034	42559	42033	42007	42035	44011	L
LA02	41032	40727	42046	42561	42045	42207	42047	44015	L
LA03	41038	40757	42055	42562	42343	42292	42056	44018	L
LA04	41052	40710	42077	42564	42076	42004	42078	44025	L
LA05	41094	40707	42185	42567	42184	42183	42107	44001	L
LA06	41104	40713	42208	42505	42054	42206	42209	44066	L
LA07	41122	40722	42252	42571	42019	42345	42253	44028	L
LA08	41124	40739	42256	42572	42263	42255	42257	44029	L
LA09	41130	40716	42072	42574	42325	42267	42269	44032	L
LA10	41134	40721	42276	42576	42332	42275	42277	44034	L
LA11	41135	40703	42280	42580	42265	42279	42281	44035	L
LA12	41136	40752	42144	42551	42143	42268	42145	44049	L
LA13	41142	40734	42333	42577	42075	42291	42293	44038	L
LA14	41144	40733	42296	42578	42350	42295	42297	44039	L
LA15	41146	40715	42300	42579	42351	42299	42301	44040	L
LA16	41158	40743	42245	42581	42129	42200	42250	44086	L
OC30	41008	40807	42079	42552	42236	42251	42080	44026	H

Set	K	L	F	E	D	C	B	A	density
OC31	41128	40804	42060	42573	42197	42347	42061	44020	H
OC32	41018	40801	42025	42556	42362	42024	42026	44008	H
OC33	41028	40806	42040	42560	42039	42348	42041	44013	H
OC34	41102	40803	42203	42568	42027	42201	42204	44064	H
OC35	41106	40808	42213	42569	42212	42211	42214	44067	H
OC36	41110	40809	42346	42511	42349	42138	42089	44003	H
OC37	41132	40802	42272	42575	42073	42271	42273	44033	H
OC38	41138	40810	42284	42517	42003	42202	42285	44036	H
LA60	41162	40900	42231	42570	42232	42353	42233	44074	H
LA61	41160	40901	42304	42566	42303	42302	42305	44068	H
LA62	41167	40902	42167	42565	42168	42103	42169	44055	H
LA63	41059	40903	42175	42563	42176	42105	42177	44081	H
LA64	41166	40904	42094	42582	42093	42092	42108	44076	H
LA71	41010	40204	42013	42553	42360	42012	42014	44004	H
LA72	41012	40205	42016	42554	42005	42015	42361	44005	H
LA73	41016	40207	42006	42555	42096	42021	42023	44007	H
LA74	41020	40221	42028	42557	42009	42259	42029	44009	H
LA75	41022	40210	42031	42558	42010	42030	42032	44010	H

Set	K	L	F	E	D	C	B	A	density
OC40	41181	41149	40106	42502	42166	42218	42071	44079	SH
OC41	41109	41182	40107	42520	42222	42224	42382	44101	SH
OC42	41168	41183	40108	42519	42315	42317	42383	44090	SH
OC43	41147	41186	40109	42515	42258	42266	42365	44000	SH
OC44	41191	41192	40110	42514	42115	42174	42288	44097	SH
OC45	41119	41187	40111	42501	42247	42173	42260	44078	SH
OC46	41096	41161	40112	42512	42178	42195	42043	44042	SH
OC47	41081	41108	40113	42509	42308	42217	42067	44060	SH
OC48	41184	41189	40114	42518	42085	42310	42087	44069	SH
OC49	41148	41169	40115	42385	42319	42364	42321	44091	SH
OC50	41179	41180	40101	42516	42098	42264	42283	44093	SH
OC51	41005	41006	40103	42513	42070	42069	42118	44023	SH
OC52	41029	41030	40102	42510	42042	42008	42044	44014	SH
OC53	41033	41034	40118	42506	42048	42066	42050	44016	SH
OC54	41055	41056	40117	42503	42074	42081	42126	44043	SH
OC55	41065	41089	40104	42504	42221	42062	42068	44022	SH
OC56	41125	41126	40116	42508	42216	42344	42261	44030	SH
OC57	41139	41140	40105	42507	42099	42287	42289	44037	SH

L = Low density; H = High density; SH = Super High density.

Spares:

LA:	40119	40231	40718	40811	41004	41103	41116	41137	41176
	42049	42083	42095	42101	42102	42196	42294	42356	42381
	44002	44059	44083						
OO:	44024								
PM:	44100								

EAST COAST

East Coast operates 14 refurbished HST sets on the ECML. As well as serving non-electrified destinations such as Hull and Inverness the East Coast HSTs also work alongside Class 91s and Mark 4 sets on services to Leeds, Newcastle and Edinburgh. Set EC64 (ex-NL05) transferred from East Midlands Trains in 2011.

Number of sets: 14.
Maximum number of daily diagrams: 13.
Formations: 9-cars.
Allocation: Craigentinny (Edinburgh).
Other maintenance depot: Neville Hill (Leeds).
Operation: London King's Cross–Leeds/Harrogate/Skipton/Hull/Lincoln/ Newcastle/Edinburgh/Aberdeen/Inverness.

Set	M	L	J	G	F	E	D	C	B
EC51	41120	41150	40748	42215	42091	42146	42150	42154	44094
EC52	41039	41040	40735	42323	42189	42057	42058	42059	44019
EC53	41090	41044	40737	42340	42127	42063	42064	42065	44045
EC54	41087	41088	40706	42104	42161	42171	42172	42219	44056
EC55	41091	41092	40704	42179	42188	42180	42181	42106	44058
EC56	41170	41118	40720	42241	42363	42242	42243	42244	44098
EC57	41151	41152	40740	42226	42128	42182	42186	42190	44080
EC58	41097	41098	40750	42158	42238	42191	42192	42193	44061
EC59	41099	41100	40711	42235	42239	42240	42198	42199	44063
EC60	41066	41164	40742	42122	42116	42357	42134	42355	44031
EC61	41115	41165	40702	42117	42159	42160	42109	42110	44057
EC62	41185	41095	40701	42306	42326	42330	42237	42307	44075
EC63	41159	41083	40708	42163	42286	42228	42130	42322	44050
EC64	41062	41154	40732	42125	42335	42123	42205	42210	44073

Spares:

EC: 40705   41190   42354   44077

GRAND CENTRAL

GC operates HSTs between Sunderland and King's Cross. Formations are flexible: there are enough vehicles to form three 6-car sets or one set can be disbanded to enable the sets to run as 7-cars at times of high demand.

Number of sets: 3.
Maximum number of daily diagrams: 2.
Formations: 6-cars or 7-cars.
Allocation: Heaton (Newcastle).
Operation: London King's Cross–Sunderland.

Set	TF	TSB	TS	TS	TS	TF*
GC01	41201	40424	42403	42402	42401	41204
GC02	41202	40426	42406	42405	42404	41205
GC03	41203	40433	42409	42408	42407	41206

* declassified.

EAST MIDLANDS TRAINS

East Midlands Trains HSTs are concentrated on the Nottingham corridor during the day, with early morning and evening services to Leeds for servicing at Neville Hill.

Number of sets: 10.
Maximum number of daily diagrams: 9.
Formations: 8-cars.
Allocation: Neville Hill (Leeds).
Other maintenance depot: Derby Etches Park.
Operation: London St Pancras–Nottingham, Sheffield/Leeds.

Set	J	G	F	E	D	C	B	A
NL01	41057	41084	40730	42327	42111	42112	42113	44041
NL02	41112	41067	40754	42194	42229	42227	42121	44027
NL03	41061	41068	40741	42337	42119	42120	42121	44025
NL04	41077	41064	40749	42151	42164	42165	42153	44047
NL06	41156	41041	40746	42132	42131	42331	42133	44046
NL07	41111	41070	40751	42135	42136	42136	42137	44044
NL08	41071	41072	40753	42329	42139	42140	42141	44048
NL10	41075	41076	40756	42328	42341	42148	42149	44051
NL11	41117	41046	40728	42220	42100	42230	42124	44085
NL12	41079	41069	40700	42155	42156	42157	42152	44070

Spares:

NL: 41063   41113   42384   44071

CROSSCOUNTRY

CrossCountry reintroduced regular HST diagrams on its services from the December 2008 timetable. Trains run in 7-car formation with the spare TS coaches regularly used in traffic as coaches "C", "D" or "E".

Number of sets: 5.
Maximum number of daily diagrams: 4.
Formations: 7-cars.
Allocation: Craigentinny (Edinburgh).
Other maintenance depots: Laira (Plymouth) or Neville Hill (Leeds).
Operation: Edinburgh–Leeds–Plymouth is the core route with some services extending to Dundee or Penzance.

Set	A	B	C	D	E	F	G
XC01	41193	45001	42368	42369	42367	42366	44021
XC02	41194	45002	42375	42373	42374	42371	44072
XC03	41195	45003	42290	42377	42377	42376	44052
XC04	41026	45004	42038	42037	42036	42380	44012
XC05	41035	45005	42051	42053	42052	42379	44017

Spares:

EC: 42097   42234   42342   42370   42372

3. SALOONS

Several specialist passenger carrying vehicles, normally referred to as saloons are permitted to run on the National Rail system. Many of these are to pre-nationalisation designs.

WCJS FIRST CLASS SALOON

Built 1892 by LNWR, Wolverton. Originally dining saloon mounted on six-wheel bogies. Rebuilt with new underframe with four-wheel bogies in 1927. Rebuilt 1960 as observation saloon with DMU end. Gangwayed at other end. The interior has a saloon, kitchen, guards vestibule and observation lounge. 19/– 1T. Gresley bogies. 28.5 t. 75 mph.

41   (484, 45018) x   **M**   WC *WC*   CS

LNWR DINING SALOON

Built 1890 by LNWR, Wolverton. Mounted on the underframe of LMS General Utility Van 37908 in the 1980s. Contains kitchen and dining area seating 12 at tables for two. 12/–. Gresley bogies. 75 mph. 25.4 t.

159   (5159)   x   **M**   WC *WC*   CS

GNR FIRST CLASS SALOON

Built 1912 by GNR, Doncaster. Contains entrance vestibule, lavatory, two separate saloons, library and luggage space. 19/– 1T. Gresley bogies. 75 mph. 29.4 t.

Non-standard livery: Teak.

807   (4807)   x   **0**   WC *WC*   CS

LNER GENERAL MANAGERS SALOON

Built 1945 by LNER, York. Gangwayed at one end with a veranda at the other. The interior has a dining saloon seating 12, kitchen, toilet, office and nine seat lounge. 21/– 1T. B4 bogies. 75 mph. ETH 3. 35.7 t.

1999   (902260)   **M**   BE   CS   DINING CAR No. 2

GENERAL MANAGER'S SALOON

Renumbered 1989 from London Midland Region departmental series. Formerly the LMR General Manager's saloon. Rebuilt from LMS period 1 Corridor Brake First M5033M to dia 1654 and mounted on the underframe of BR suburban Brake Standard M43232. Screw couplings have been removed. B4 bogies. 100 mph. ETH 2X.

LMS Lot No. 326 Derby 1927. 27.5 t.

6320   (5033, DM 395707)   x   **M**   62   *62*   SK

BELMOND BRITISH PULLMAN SUPPORT CAR

Converted 199? from Courier vehicle converted from Mark 1 Corridor Brake Standard 1986–87. Toilet retained and former compartment area replaced with train manager's office, crew locker room, linen store and dry goods store. The former luggage area has been adapted for use as an engineers' compartment and workshop. Commonwealth bogies. 100 mph. ETH 2.

Lot No. 30721 Wolverton 1963. 37 t.

99545   (35466, 80207)   **PC**   BE   *BP*   SL   BAGGAGE CAR No. 11

SERVICE CAR

Converted from BR Mark 1 Corridor Brake Standard. Commonwealth bogies. 100 mph. ETH 2.

Lot No. 30721 Wolverton 1963.

99886   (35407)   x   **M** WC   *WC*   CS   SERVICE CAR No. 86

ROYAL SCOTSMAN SALOONS

Built 1960 by Metro-Cammell as Pullman Kitchen First for East Coast Main Line services. Rebuilt 2013 as dining car. Commonwealth bogies. 38.5 t.

99960   (321 SWIFT)   **M**   BE   *RS*   CS   DINING CAR No. 2

Built 1960 by Metro-Cammell as Pullman Parlour First (§ Pullman Kitchen First) for East Coast Main Line services. Rebuilt 1990 as sleeping cars with four twin sleeping rooms (*§ three twin sleeping rooms and two single sleeping rooms at each end). Commonwealth bogies. 38.5 t.

99961	(324 AMBER) *	**M**	BE	*RS*	CS	STATE CAR 1
99962	(329 PEARL)	**M**	BE	*RS*	CS	STATE CAR 2
99963	(331 TOPAZ)	**M**	BE	*RS*	CS	STATE CAR 3
99964	(313 FINCH) §	**M**	BE	*RS*	CS	STATE CAR 4

Built 1960 by Metro-Cammell as Pullman Kitchen First for East Coast Main Line services. Rebuilt 1990 as observation car with open verandah seating 32. Commonwealth bogies. 38.5 t.

| 99965 (319 SNIPE) | **M** | BE | *RS* | CS | OBSERVATION CAR |

Built 1960 by Metro-Cammell as Pullman Kitchen First for East Coast Main Line services. Rebuilt 1993 as dining car. Commonwealth bogies. 38.5 t.

| 99967 (317 RAVEN) | **M** | BE | *RS* | CS | DINING CAR No. 1 |

Mark 3A. Converted 1997 from a Sleeping Car at Carnforth Railway Restoration & Engineering Services. BT10 bogies. Attendant's and adjacent two sleeping compartments converted to generator room containing a 160 kW Volvo unit. In 99968 four sleeping compartments remain for staff use with another converted for use as a staff shower and toilet. The remaining five sleeping compartments have been replaced by two passenger cabins. In 99969 seven sleeping compartments remain for staff use. A further sleeping compartment, along with one toilet, have been converted to store rooms. The other two sleeping compartments have been combined to form a crew mess. ETH 7X. 41.5 t.

Lot No. 30960 Derby 1981–83.

| 99968 (10541) | **M** | BE | *RS* | CS | STATE CAR 5 |
| 99969 (10556) | **M** | BE | *RS* | CS | SERVICE CAR |

RAILFILMS "LMS CLUB CAR"

Converted from BR Mark 1 Open Standard at Carnforth Railway Restoration & Engineering Services in 1994. Contains kitchen, pantry and two dining saloons. 20/– 1T. Commonwealth bogies. 100 mph. ETH 4.

Lot No. 30724 York 1963. 37 t.

| 99993 (5067) | x | **M** | RA | *ST* | CS | LMS CLUB CAR |

BR INSPECTION SALOON

Mark 1. Short frames. Non-gangwayed. Observation windows at each end. The interior layout consists of two saloons interspersed by a central lavatory/kitchen/guards/luggage section. 90 mph.

BR Wagon Lot No. 3095 Swindon 1957. B4 bogies. 30.5 t.

| 999506 | **M** | WC | *WC* | CS | |

4. PULLMAN CAR COMPANY SERIES

Pullman cars have never generally been numbered as such, although many have carried numbers, instead they have carried titles. However, a scheme of schedule numbers exists which generally lists cars in chronological order. In this section those numbers are shown followed by the car's title. Cars described as "kitchen" contain a kitchen in addition to passenger accommodation and have gas cooking unless otherwise stated. Cars described as "parlour" consist entirely of passenger accommodation. Cars described as "brake" contain a compartment for the use of the guard and a luggage compartment in addition to passenger accommodation.

PULLMAN PARLOUR FIRST

Built 1927 by Midland Carriage & Wagon Company. 26/– 2T. Gresley bogies. ETH 2. 41 t.

| 213 | MINERVA | PC | BE | BP | SL |

PULLMAN PARLOUR FIRST

Built 1928 by Metropolitan Carriage & Wagon Company. 24/– 2T. Gresley bogies. ETH 4. 40 t.

| 239 | AGATHA | PC | BE | | SL |
| 243 | LUCILLE | PC | BE | BP | SL |

PULLMAN KITCHEN FIRST

Built 1925 by BRCW. Rebuilt by Midland Carriage & Wagon Company in 1928. 20/– 1T. Gresley bogies. ETH 4. 41 t.

| 245 | IBIS | PC | BE | BP | SL |

PULLMAN PARLOUR FIRST

Built 1928 by Metropolitan Carriage & Wagon Company. 24/– 2T. Gresley bogies. ETH 4.

| 254 | ZENA | PC | BE | BP | SL |

PULLMAN KITCHEN FIRST

Built 1928 by Metropolitan Carriage & Wagon Company. 20/– 1T. Gresley bogies. ETH 4. 42 t.

| 255 | IONE | PC | BE | BP | SL |

PULLMAN KITCHEN COMPOSITE

Built 1932 by Metropolitan Carriage & Wagon Company. Originally included in 6-Pul EMU. Electric cooking. 12/16 1T. EMU bogies.

264	RUTH	**PC**	BE		SL

PULLMAN KITCHEN FIRST

Built 1932 by Metropolitan Carriage & Wagon Company. Originally included in "Brighton Belle" EMUs but now used as hauled stock. Electric cooking. 20/– 1T. B5 (SR) bogies (§ EMU bogies). ETH 2. 44 t.

280	AUDREY		**PC**	BE	*BP*	SL
281	GWEN		**PC**	BE	*BP*	SL
283	MONA	§	**PC**	BE		SL
284	VERA		**PC**	BE	*BP*	SL

PULLMAN PARLOUR THIRD

Built 1932 by Metropolitan Carriage & Wagon Company. Originally included in "Brighton Belle" EMUs. –/56 2T. EMU bogies.

Non-standard livery: BR Revised Pullman (blue & white lined out in white).

286	CAR No. 86	**O**	BE		SL

PULLMAN BRAKE THIRD

Built 1932 by Metropolitan Carriage & Wagon Company. Originally driving motor cars in "Brighton Belle" EMUs. Traction and control equipment removed for use as hauled stock. –/48 1T. EMU bogies.

292	CAR No. 92	**PC**	BE		SL
293	CAR No. 93	**PC**	BE		SL

PULLMAN PARLOUR FIRST

Built 1951 by Birmingham Railway Carriage & Wagon Company. 32/– 2T. Gresley bogies. ETH 3. 39 t.

301	PERSEUS	**PC**	BE	*BP*	SL

Built 1952 by Pullman Car Company, Preston Park using underframe and bogies from 176 RAINBOW, the body of which had been destroyed by fire. 26/– 2T. Gresley bogies. ETH 4. 38 t.

302	PHOENIX	**PC**	BE	*BP*	SL

PULLMAN PARLOUR FIRST

Built 1951 by Birmingham Railway Carriage & Wagon Company. 32/– 2T.
Gresley bogies. ETH 3. 39 t.

| 308 | CYGNUS | **PC** | BE | *BP* | SL |

PULLMAN BAR FIRST

Built 1951 by Birmingham Railway Carriage & Wagon Company. Rebuilt
1999 by Blake Fabrications, Edinburgh with original timber-framed body
replaced by a new fabricated steel body. Contains kitchen, bar, dining
saloon and coupé. Electric cooking. 14/– 1T. Gresley bogies. ETH 3.

| 310 | PEGASUS | x | **PC** | RA | *ST* | CS |

Also carries "THE TRIANON BAR" branding.

PULLMAN PALOUR FIRST

Built 1960–61 by Metro-Cammell for East Coast Main Line services. –/36 2T.
Commonwealth bogies. 38.5 t.

| 326 | EMERALD | x | **PC** | WC | *WC* | CS |

PULLMAN KITCHEN SECOND

Built 1960–61 by Metro-Cammell for East Coast Main Line services.
Commonwealth bogies. –/30 1T. 40 t.

| 335 | CAR No. 335 | x | **PC** | VT | *VT* | TM |

PULLMAN PARLOUR SECOND

Built 1960–61 by Metro-Cammell for East Coast Main Line services. 347, 348
and 350 are used as Open Firsts. –/42 2T. Commonwealth bogies. 38.5 t.

347	CAR No. 347	x	**M**	WC	*WC*	CS	
348	CAR No. 348	x	**M**	WC	*WC*	CS	
349	CAR No. 349	x	**PC**	VT	*VT*	TM	
350	CAR No. 350	x	**M**	WC	*WC*	CS	
352	CAR No. 352	x	**PC**	WC	*WC*	CS	AMETHYST
353	CAR No. 353	x	**PC**	VT	*VT*	TM	

PULLMAN SECOND BAR

Built 1960–61 by Metro-Cammell for East Coast Main Line services.
–/24+17 bar seats. Commonwealth bogies. 38.5 t.

| 354 | THE HADRIAN BAR | x | **PC** | WC | *WC* | CS |

5. LOCOMOTIVE SUPPORT COACHES

These carriages have been adapted from Mark 1s and Mark 2s for use as support coaches for heritage steam and diesel locomotives. Some seating is retained for the use of personnel supporting the locomotives operation with the remainder of the carriage adapted for storage, workshop, dormitory and catering purposes. These carriages can spend considerable periods of time off the national railway system when the locomotives they support are not being used on that system. No owner or operator details are included in this section. After the depot code, the locomotive(s) each carriage is usually used to support is given.

CORRIDOR BRAKE FIRST

Mark 1. Commonwealth bogies. ETH 2.

14007. Lot No. 30382 Swindon 1959. 35 t.
17015. Lot No. 30668 Swindon 1961. 36 t.
17025. Lot No. 30718 Swindon 1963. Metal window frames. 36 t.

14007	(14007, 17007)	x	**M**	NY	LNER 61264
17015	(14015)		**CC**	TM	Tyseley Locomotive Works-based locos
17025	(14025)	v	**M**	CS	LMS 45690

CORRIDOR BRAKE FIRST

Mark 2A. Pressure ventilated. B4 bogies. ETH 4.

14064. Lot No. 30775 Derby 1967–68. 32 t.
14099/17096. Lot No. 30786 Derby 1968. 32 t.

14064	(14064, 17064)	x	**M**	CS	LMS 45305/BR 70013
14099	(14099, 17099)	v	**M**	CS	LMS 45305/BR 70013
17096	(14096)		**PC**	SL	SR 35028 MERCATOR

CORRIDOR BRAKE COMPOSITE

Mark 1. ETH 2.

21096. Lot No. 30185 Metro-Cammell 1956. BR Mark 1 bogies. 32.5 t.
21232. Lot No. 30574 GRCW 1960. B4 bogies. 34 t.
21249. Lot No. 30669 Swindon 1961–62. Commonwealth bogies. 36 t.

21096	x	**M**	NY	LNER 60007
21232	x	**M**	SK	LMS 46233
21249	x	**CC**	BH	New Build 60163

CORRIDOR BRAKE STANDARD

Mark 1. Metal window frames and melamine interior panelling. ETH 2.

35317–322. Lot No. 30699 Wolverton 1962–63. Commonwealth bogies. 37 t.
35449. Lot No. 30728 Wolverton 1963. Commonwealth bogies. 37 t.
35451–486. Lot No. 30721 Wolverton 1963. Commonwealth bogies. 37 t.

35317	x	**CC**	SH	LNER 60019
35322	x	**M**	CS	WCRC Carnforth-based locomotives
35449	x	**M**	BQ	LMS 45231
35451	x	**CC**	SH	SR 34046
35461	x	**CH**	SH	GWR 5029
35463	v	**M**	CS	WCRC Carnforth-based locomotives
35465	x	**CC**	SH	BR 70000
35468	x	**M**	YK	National Railway Museum locomotives
35470	v	**CH**	TM	Tyseley Locomotive Works-based locos
35476	x	**M**	SK	LMS 46233
35486	x	**M**	TN	LNER 60009/61994

CORRIDOR BRAKE FIRST

Mark 2C. Pressure ventilated. Renumbered when declassified. B4 bogies. ETH 4.

Lot No. 30796 Derby 1969–70. 32.5 t.

35508 (14128, 17128)	**M**	BQ	LMS 44871/45407

CORRIDOR BRAKE FIRST

Mark 2A. Pressure ventilated. Renumbered when declassified. B4 bogies. ETH 4.

Lot No. 30786 Derby 1968. 32 t.

35517 (14088, 17088)	b	**M**	BQ	LMS 44871/45407
35518 (14097, 17097)	b	**G**	SH	SR 34067

COURIER VEHICLE

Mark 1. Converted 1986–87 from Corridor Brake Standards. ETH 2.

80204/217. Lot No. 30699 Wolverton 1962. Commonwealth bogies. 37 t.
80220. Lot No. 30573 Gloucester 1960. B4 bogies. 33 t.

80204 (35297)	**M**	CS	WCRC Carnforth-based locomotives
80217 (35299)	**M**	CS	WCRC Carnforth-based locomotives
80220 (35276)	**M**	NY	LNER 62005

6. 95xxx & 99xxx RANGE NUMBER CONVERSION TABLE

The following table is presented to help readers identify vehicles which may still carry numbers in the 95xxx and 99xxx number ranges of the former private owner number series, which is no longer in general use.

9xxxx	BR No.	9xxxx	BR No.	9xxxx	BR No.
95402	Pullman 326	99349	Pullman 349	99672	549
99035	35322	99350	Pullman 350	99673	550
99040	21232	99352	Pullman 352	99674	551
99041	35476	99353	Pullman 353	99675	552
99052	Saloon 41	99354	Pullman 354	99676	553
99121	3105	99361	Pullman 335	99677	586
99122	3106	99371	3128	99678	504
99125	3113	99405	35486	99679	506
99127	3117	99530	Pullman 301	99680	17102
99128	3130	99531	Pullman 302	99710	18767
99131	Saloon 1999	99532	Pullman 308	99716 *	18808
99241	35449	99534	Pullman 245	99718	18862
99302	13323	99535	Pullman 213	99721	18756
99304	21256	99536	Pullman 254	99723	35459
99311	1882	99537	Pullman 280	99880	Saloon 159
99312	35463	99539	Pullman 255	99881	Saloon 807
99317	3766	99541	Pullman 243	99887	2127
99319	17168	99543	Pullman 284	99953	35468
99326	4954	99546	Pullman 281	99966	34525
99327	5044	99547	Pullman 292	99970	Pullman 232
99328	5033	99548	Pullman 293	99972	Pullman 318
99329	4931	99670	546	99973	324
99347	Pullman 347	99671	548	99974	Pullman 328
99348	Pullman 348				

* The number 99716 has also been applied to 3416 for filming purposes.

▲ First Group Dynamic Lines-liveried Mark 3B Open Brake Unclassified 17174 at Exeter St Davids on 26/07/14. **Stewart Armstrong**

▼ BR carmine & cream-liveried Mark 1 Corridor Brake Generator Standard 35469 at Carluke on 21/06/14. **Robin Ralston**

▲ Abellio Greater Anglia-liveried Mark 3B Driving Brake Van 82152 arrives at Colchester leading the 10.00 London Liverpool Street–Norwich on 15/04/14.
Antony Guppy

▼ Stagecoach East Midlands Trains-liveried HST Trailer First 41067 at Nottingham on 03/08/14.
Robert Pritchard

▲ Grand Central-liveried HST Trailer Standard 42402 at Hitchin on 09/05/14.
Mark Beal

▼ First Great Western Dynamic Lines-liveried HST Trailer Standard 42561 (converted in 2014 from Trailer First 41031) at Newport on 29/07/14.
Stewart Armstrong

▲ BR maroon-liveried First Class Queen of Scots Saloon No. 41 at Cleghorn on 30/09/13 on the rear of a Carnforth–Glenfinnan charter. **Robin Ralston**

▼ Royal Scotsman Saloon 99964 "STATE CAR 4" at Kinneil on 15/04/14. **Ian Lothian**

▲ Pullman Car Company-liveried Pullman Parlour First 254 "ZENA" at Newton Abbot on 25/04/14. **Tony Christie**

▼ Porterbrook-liveried EMU Translator Vehicle 6376 at Wimbledon depot on 30/06/14. **Brian Garvin**

▲ Network Rail Driving Trailer Coach 9701 leads a Perth–Stirling test train (powered by 37601) on the approaches to Alloa on 10/04/14. **Ian Lothian**

▼ Network Rail Ultrasonic Test Coach 62384 (converted from a Class 421 EMU MBSO) at Torquay on 17/06/14. **Tony Christie**

▲ BR Southern Region green-liveried Inspection Saloon 975025 "CAROLINE" passes Loughborough, being propelled by 37409, on 14/04/12.　　**Paul Biggs**

▼ Overhead Line Equipment Test Coach 975091 "MENTOR" (converted from a Mark 1 Corridor Brake Standard) at Totnes on 25/07/14.　　**Tony Christie**

▲ New Measurement Train Lecture Coach 975984 (converted from a prototype HST catering vehicle) at Stoke-on-Trent on 07/08/14. **Robert Pritchard**

▼ NMT Overhead Line Equipment Test Coast 977993 (converted from an HST TGS and fitted with a pantograph) at Cleghorn on 10/09/13. **Robin Ralston**

7. SET FORMATIONS

CHILTERN RAILWAYS MARK 3 SET FORMATIONS

Chiltern Railways uses a number of loco-hauled rakes on services principally between London Marylebone and Birmingham Moor Street. Rake AL06 is used on a commuter train between Marylebone and Banbury. All coaches apart from rake AL06 and spare 12094 have been rebuilt at Wabtec, Doncaster and fitted with sliding plug doors.

Set							DVT
AL01	12610	12613	12614	12615	12602	10273	82301
AL02	12603	12606	12607	12608	12609	10272	82302
AL03	12616	12617	12618	12619	12604	10274	82303
AL04	12623	12605	12621	12627	12625	10271	82304
AL06	12043	12119	12017	11029	11031	12054	82305

| Spare | 12094 | 12620 | 82309 |

EAST COAST MARK 4 SET FORMATIONS

The East Coast Mark 4 sets generally run in fixed formations since their refurbishment at Bombardier, Wakefield in 2003–05. These rakes are listed below. Class 91 locomotives are positioned next to Coach B.

Set	B	C	D	E	F	H	K	L	M	DVT
BN01	12207	12417	12415	12414	12307	10307	11298	11301	11401	82207
BN02	12232	12402	12450	12448	12302	10302	11299	11302	11402	82202
BN03	12201	12401	12459	12478	12301	10320	11277	11303	11403	82219
BN04	12202	12480	12421	12518	12327	10303	11278	11304	11404	82209
BN05	12209	12486	12520	12522	12300	10326	11219	11305	11405	82210
BN06	12208	12406	12420	12422	12313	10309	11279	11306	11406	82208
BN07	12231	12411	12405	12489	12329	10323	11280	11307	11407	82204
BN08	12205	12481	12485	12407	12328	10300	11229	11308	11408	82211
BN09	12230	12513	12483	12514	12308	10304	11281	11309	11409	82215
BN10	12214	12419	12488	12443	12305	10331	11282	11310	11410	82205
BN11	12203	12437	12436	12484	12315	10308	11283	11311	11411	82218
BN12	12212	12431	12404	12426	12330	10333	11284	11312	11412	82212
BN13	12228	12469	12424	12411	12311	10313	11285	11313	11413	82213
BN14	12229	12410	12526	12423	12312	10332	11201	11314	11414	82206
BN15	12226	12442	12409	12515	12309	10306	11286	11315	11415	82214
BN16	12213	12428	12445	12433	12304	10315	11287	11316	11416	82222
BN17	12223	12444	12427	12432	12303	10324	11288	11317	11417	82225
BN18	12215	12453	12468	12467	12324	10305	11289	11318	11418	82220
BN19	12211	12434	12400	12470	12310	10318	11290	11319	11419	82201
BN20	12224	12477	12439	12440	12326	10321	11241	11320	11420	82200
BN21	12222	12461	12441	12476	12323	10330	11244	11321	11421	82227
BN22	12210	12452	12460	12473	12316	10301	11291	11322	11422	82230
BN23	12225	12454	12456	12455	12318	10325	11292	11323	11423	82226
BN24	12219	12447	12425	12403	12319	10328	11293	11324	11424	82229
BN25	12217	12446	12519	12464	12322	10312	11294	11325	11425	82216
BN26	12220	12474	12465	12429	12325	10311	11295	11326	11426	82223
BN27	12216	12449	12466	12538	12317	10319	11237	11327	11427	82228
BN28	12218	12458	12463	12533	12320	10310	11273	11328	11428	82217
BN29	12204	12462	12457	12438	12321	10317	11998	11329	11429	82231
BN30	12227	12471	12534	12472	12331	10329	11999	11330	11430	82203
Spare	12200									82224

8. SERVICE STOCK

Vehicles in this section are used for internal purposes within the railway industry, ie they do not generate revenue from outside the industry. Most are numbered in the former BR departmental number series.

BARRIER, ESCORT & TRANSLATOR VEHICLES

These vehicles are used to move multiple unit, HST and other vehicles around the national railway system.

HST Barrier Vehicles. Mark 1/2A. Renumbered from BR departmental series, or converted from various types. B4 bogies (* Commonwealth bogies).

Non-standard livery: 6340, 6344, 6346 All over dark blue.

6330. Mark 2A. Lot No. 30786 Derby 1968.
6336/38/44. Mark 1. Lot No. 30715 Gloucester 1962.
6340. Mark 1. Lot No. 30669 Swindon 1962.
6346. Mark 2A. Lot No. 30777 Derby 1967.
6348. Mark 1. Lot No. 30163 Pressed Steel 1957.

6330	(14084, 975629)		**FB**	A	*GW*	LA
6336	(81591, 92185)		**FB**	A	*GW*	LA
6338	(81581, 92180)		**FB**	A	*GW*	LA
6340	(21251, 975678)	*	**0**	A	*EC*	EC
6344	(81263, 92080)		**0**	A	*EC*	EC
6346	(9422)		**0**	A	*EC*	EC
6348	(81233, 92963)		**FB**	A	*GW*	LA

Mark 4 Barrier Vehicles. Mark 2A/2C. Converted from Corridor First (*) or Open Brake Standard. B4 bogies.

Non-standard livery: 6358 All over dark blue.

6352/53. Mark 2A. Lot No. 30774 Derby 1968.
6354/55. Mark 2C. Lot No. 30820 Derby 1970.
6358/59. Mark 2A. Lot No. 30788 Derby 1968.

6352	(13465, 19465)	*	**GN**	E	*EC*	BN
6353	(13478, 19478)	*	**GN**	E	*EC*	BN
6354	(9459)		**GN**	E	*EC*	BN
6355	(9477)		**GN**	E	*EC*	BN
6358	(9432)		**0**	E	*EC*	BN
6359	(9429)		**GN**	E	*EC*	BN

EMU Translator Vehicles. Mark 1. Converted 1980 from Restaurant Unclassified Opens. Commonwealth bogies.

Lot No. 30647 Wolverton 1959–61.

6376	(1021, 975973)	**PB**	P	*CS*	RU *(works with 6377)*
6377	(1042, 975975)	**PB**	P	*CS*	RU *(works with 6376)*
6378	(1054, 975971)	**PB**	P	*CS*	RU *(works with 6379)*
6379	(1059, 975972)	**PB**	P	*CS*	RU *(works with 6378)*

HST Barrier Vehicles. Mark 1. Converted from Gangwayed Brake Vans in 1994–95. B4 bogies.

6392. Lot No. 30715 Gloucester 1962.
6393/97. Lot No. 30716 Gloucester 1962.
6394. Lot No. 30162 Pressed Steel 1956–57.
6398/99. Lot No. 30400 Pressed Steel 1957–58.

6392	(81588, 92183)	**PB**	P		LM
6393	(81609, 92196)	**PB**	P	*EC*	EC
6394	(80878, 92906)	**P**	P	*EC*	EC
6397	(81600, 92190)	**PB**	P		LM
6398	(81471, 92126)	**PB**	EM	*EM*	NL
6399	(81367, 92994)	**PB**	EM	*EM*	NL

Escort Coaches. Converted from Mark 2A Open Brake Standards. These vehicles use the same bodyshell as the Mark 2A Corridor Brake First. B4 bogies.

9419. Lot No.30777 Derby 1970.
9428. Lot No.30820 Derby 1970.

9419	**DS**	DR	*DR*	KM
9428	**DS**	DR	*DR*	KM

EMU Translator Vehicles. Converted from Class 508 driving cars.

64664. Lot No. 30979 York 1979–80.
64707. Lot No. 30981 York 1979–80.

64664	**AG**	A	*GB*	PG	*(works with 64707)*
64707	**AG**	A	*GB*	PG	*(works with 64664)*

Eurostar Barrier Vehicles. Mark 1. Converted from General Utility Vans with bodies removed. Fitted with B4 bogies for use as Eurostar barrier vehicles.

96380/381. Lot No. 30417 Pressed Steel 1958–59.
96383. Lot No. 30565 Pressed Steel 1959.
96384. Lot No. 30616 Pressed Steel 1959–60.

96380	(86386, 6380)	**B**	EU	*EU*	TI
96381	(86187, 6381)	**B**	EU	*EU*	TI
96383	(86664, 6383)	**B**	EU	*EU*	TI
96384	(86955, 6384)	**B**	EU	*EU*	TI

EMU Translator Vehicles. Converted from various Mark 1s.

Non-standard livery: All over blue.

975864. Lot No. 30054 Eastleigh 1951–54. Commonwealth bogies.
975867. Lot No. 30014 York 1950–51. Commonwealth bogies.
975875. Lot No. 30143 Charles Roberts 1954–55. Commonwealth bogies.
975974/978. Lot No. 30647 Wolverton 1959–61. B4 bogies.
977087. Lot No. 30229 Metro–Cammell 1955–57. Commonwealth bogies.

975864	(3849)	**HB**	E	*GB*	PG	*(works with 975867)*
975867	(1006)	**HB**	E	*GB*	PG	*(works with 975864)*
975875	(34643)	**0**	E	*GB*	PG	*(works with 977087)*

975974	(1030)	**AG**	A	*GB*	PG	Paschar	*(works with 975978)*
975978	(1025)	**AG**	A	*GB*	PG	Perpetiel	*(works with 975974)*
977087	(34971)	**0**	E	*GB*	PG		*(works with 975875)*

LABORATORY, TESTING & INSPECTION COACHES

These coaches are used for research, development, testing and inspection on the national railway system. Many are fitted with sophisticated technical equipment.

Plain Line Pattern Recognition Coaches. Converted from BR Mark 2F Buffet First (*) or Open Standard. B4 bogies.

1256. Lot No. 30845 Derby 1973.
5981. Lot No. 30860 Derby 1973–74.

1256	(3296)	*	**Y**	NR	*DB*	ZA
5981			**Y**	NR	*DB*	ZA

Generator Vans. Mark 1. Converted from BR Mark 1 Gangwayed Brake Vans. B5 bogies.

6260. Lot No. 30400 Pressed Steel 1957–58.
6261. Lot No. 30323 Pressed Steel 1957.
6262. Lot No. 30228 Metro-Cammell 1957–58.
6263. Lot No. 30163 Pressed Steel 1957.
6264. Lot No. 30173 York 1956.

6260	(81450, 92116)	**NR** NR			ZA
6261	(81284, 92988)	**Y**	NR	*DB*	ZA
6262	(81064, 92928)	**Y**	NR	*DB*	ZA
6263	(81231, 92961)	**Y**	NR	*DB*	ZA
6264	(80971, 92923)	**Y**	NR	*DB*	ZA

Staff Coaches. Mark 2D/2F. Converted from BR Mark 2D/2F Open Brake Standard. B4 bogies.

9481. Lot No. 30824 Derby 1971.
9516/23. Lot No. 30861 Derby 1974.

9481	**Y**	NR	*DB*	ZA
9516	**Y**	NR	*DB*	ZA
9523	**Y**	NR	*DB*	ZA

Driving Trailer Coaches. Converted 2008 at Serco, Derby from Mark 2F Driving Open Brake Standards. Fitted with generator and modified to work in Blue Star push-pull mode. Disc brakes. B4 bogies.

9701–08. Lot No. 30861 Derby 1974. Converted to Driving Open Brake Standard Glasgow 1974.
9714. Lot No. 30861 Derby 1974. Converted to Driving Open Brake Standard Glasgow 1986.

9701	(9528)	**Y**	NR	*DB*	ZA
9702	(9510)	**Y**	NR	*DB*	ZA
9703	(9517)	**Y**	NR	*DB*	ZA

| 9708 | (9530) | Y | NR | *DB* | ZA |
| 9714 | (9536) | Y | NR | *DB* | ZA |

Ultrasonic Test Coach. Converted from Class 421 EMU MBSO.

62287. Lot No. 30808. York 1970. SR Mark 6 bogies.
62384. Lot No. 30816. York 1970. SR Mark 6 bogies.

| 62287 | Y | NR | *DB* | ZA |
| 62384 | Y | NR | *DB* | ZA |

Test Train Brake Force Runners. Converted from Mark 2F Open Standard converted to Class 488/3 EMU TSOLH. These vehicles are included in test trains to provide brake force and are not used for any other purposes. Lot No. 30860 Derby 1973–74. B4 bogies.

| 72612 | (6156) | Y | NR | *DB* | ZA |
| 72616 | (6007) | Y | NR | *DB* | ZA |

Structure Gauging Train Coach. Converted from Mark 2F Open Standard converted to Class 488/3 EMU TSOLH. Lot No. 30860 Derby 1973–74. B4 bogies.

| 72630 | (6094) | Y | NR | *DB* | ZA *(works with 99666)* |

Plain Line Pattern Recognition Coaches. Converted from BR Mark 2F Open Standard converted to Class 488/3 EMU TSOLH. Lot No. 30860 Derby 1973–74. B4 bogies.

| 72631 | (6096) | Y | NR | *DB* | ZA |
| 72639 | (6070) | Y | NR | *DB* | ZA |

Driving Trailer Coaches. Converted from Mark 3B 110 mph Driving Brake Vans. Fitted with diesel generator. Lot No. 31042 Derby 1988.

82111	Y	NR	*DB*	ZA
82124	Y	NR		ZA
82129	Y	NR	*DB*	ZA
82145	Y	NR	*DB*	ZA

Structure Gauging Train Coach. Converted from BR Mark 2E Open First then converted to exhibition van. Lot No. 30843 Derby 1972–73. B4 bogies.

| 99666 | (3250) | Y | NR | *DB* | ZA *(works with 72630)* |

Inspection Saloon. Converted from Class 202 DEMU TRB at Stewarts Lane for use as a BR Southern Region General Manager's Saloon. Overhauled at FM Rail, Derby in 2004–05 for use as a New Trains Project Saloon. Can be used in push-pull mode with suitably equipped locomotives. Eastleigh 1958. SR Mark 4 bogies.

| 975025 | (60755) | G | NR | *DB* | ZA | CAROLINE |

Overhead Line Equipment Test Coach ("MENTOR"). Converted from BR Mark 1 Corridor Brake Standard. Lot No. 30142 Gloucester 1954–55. Fitted with pantograph. B4 bogies.

| 975091 | (34615) | Y | NR | *DB* | ZA |

New Measurement Train Conference Coach. Converted from prototype HST TF Lot No. 30848 Derby 1972. BT10 bogies.

975814 (11000, 41000) **Y** NR *DB* EC

New Measurement Train Lecture Coach. Converted from prototype HST catering vehicle. Lot No. 30849 Derby 1972–73. BT10 bogies.

975984 (10000, 40000) **Y** NR *DB* EC

Radio Survey Coach. Converted from BR Mark 2E Open Standard. Lot No. 30844 Derby 1972–73. B4 bogies.

977868 (5846) **Y** NR *DB* ZA

Staff Coach. Converted from Royal Household couchette Lot No. 30889, which in turn had been converted from BR Mark 2B Corridor Brake First. Lot No. 30790 Derby 1969. B5 bogies.

977969 (14112, 2906) **Y** NR *DB* ZA

Track Inspection Train Coach. Converted from BR Mark 2E Open Standard. Lot No. 30844 Derby 1972–73. B4 bogies.

977974 (5854) **Y** NR *DB* ZA

Electrification Measurement Coach. Converted from BR Mark 2F Open First converted to Class 488/2 EMU TFOH. Lot No. 30859 Derby 1973–74. B4 bogies.

977983 (3407, 72503) **Y** NR *DB* ZA

New Measurement Train Staff Coach. Converted from HST catering vehicle. Lot No. 30884 Derby 1976–77. BT10 bogies.

977984 (40501) **Y** P *DB* EC

Structure Gauging Train Coaches. Converted from Mark 2F Open Standard converted to Class 488/3 EMU TSOLH or from BR Mark 2D Open First subsequently declassified to Open Standard and then converted to exhibition van. B4 bogies.
977985. Lot No. 30860 Derby 1973–74.
977986. Lot No. 30821 Derby 1971.

977985 (6019, 72715) **Y** NR *DB* ZA *(works with 977986)*
977986 (3189, 99664) **Y** NR *DB* ZA *(works with 977985)*

New Measurement Train Overhead Line Equipment Test Coach. Converted from HST TGS. Lot No. 30949 Derby 1982. Fitted with pantograph. BT10 bogies.

977993 (44053) **Y** P *DB* EC

New Measurement Train Track Recording Coach. Converted from HST TGS. Lot No. 30949 Derby 1982. BT10 bogies.

977994 (44087) **Y** P *DB* EC

New Measurement Train Coach. Converted from HST catering vehicle. Lot No. 30921 Derby 1978–79. BT10 bogies. Fitted with generator.

977995 (40719, 40619) **Y** P *DB* EC

Radio Survey Coach. Converted from Mark 2F Open Standard converted to Class 488/3 EMU TSOLH. Lot No. 30860 Derby 1973–74.

977997	(72613, 6126)	Y	NR	*DB*	ZA

Track Recording Coach. Purpose built Mark 2. B4 bogies.

999550	Y	NR	*DB*	ZA

Ultrasonic Test Coaches. Converted from Class 421 EMU MBSO and Class 432 EMU MSO.

999602/605. Lot No. 30862 York 1974. SR Mk 6 bogies.
999606. Lot No. 30816. York 1970. SR Mk 6 bogies.

999602	(62483)	Y	NR	*DB*	ZA
999605	(62482)	Y	NR		ZA
999606	(62356)	Y	NR	*DB*	ZA

BREAKDOWN TRAIN COACHES

These coaches are formed in trains used for the recovery of derailed railway vehicles and were converted from BR Mark 1 Corridor Brake Standard and General Utility Van. The current use of each vehicle is given.

971001/003/004. Lot No. 30403 York/Glasgow 1958–60. Commonwealth bogies.
971002. Lot No. 30417 Pressed Steel 1958–59. Commonwealth bogies.
975087. Lot No. 30032 Wolverton 1951–52. BR Mark 1 bogies.
975464. Lot No. 30386 Charles Roberts 1956–58. Commonwealth bogies.
975471. Lot No. 30095 Wolverton 1953–55. Commonwealth bogies.
975477. Lot No. 30233 GRCW 1955–57. BR Mark 1 bogies.
975486. Lot No. 30025 Wolverton 1950–52. Commonwealth bogies.

971001	(86560, 94150)	Y	NR	*DB*	BS	Tool & Generator Van
971002	(86624, 94190)	Y	NR	*DB*	SP	Tool Van
971003	(86596, 94191)	Y	NR	*DB*	BS	Tool Van
971004	(86194, 94168)	Y	NR	*DB*	KY	Tool Van
975087	(34289)	Y	NR	*DB*	KY	Tool & Generator Van
975464	(35171)	Y	NR	*DB*	SP	Staff Coach
975471	(34543)	Y	NR	*DB*	BS	Staff Coach
975477	(35108)	Y	NR	*DB*	KY	Staff Coach
975486	(34100)	Y	NR	*DB*	SP	Tool & Generator Van

INFRASTRUCTURE MAINTENANCE COACHES

De-Icing Coaches

These coaches are used for removing ice from the conductor rail of DC lines. They were converted from Class 489 DMLVs that had originally been Class 414/3 DMBSOs.
Lot No. 30452 Ashford/Eastleigh 1959. Mk 4 bogies.

68501	(61281)	Y	NR	*GB*	TW
68504	(61286)	Y	NR	*GB*	TW
68505	(61299)	Y	NR	*GB*	TW

Winterisation Train Coach. Converted from BR Mark 2E Open Standard. Lot No. 30844 Derby 1972–73. B4 bogies.

977869 (5858)　　　　　　　**Y**　　NR　　*DR*　　　　　Edinburgh Slateford

INTERNAL USER VEHICLES

These vehicles are confined to yards and depots or do not normally move at all. Details are given of the internal user number (if allocated), type, former identity, current use and location. Many no longer see regular use.

975403 carries its original number 4598.

041379	LMS CCT 35527	Stores van	Leeman Road EY, York
041947	BR GUV 93425	Stores van	IL
042154	BR GUV 93975	Stores van	Ipswich Upper Yard
061061	BR CCT 94135	Stores van	Oxford Station
061223	BR GUV 93714	Stores van	Oxford Station
083439	BR CCT 94752	Stores van	WD
083602	BR CCT 94494	Stores van	Three Bridges Station
083637	BR NW 99203	Stores van	SL
083644	BR Ferry Van 889201	Stores van	EH
083664	BR Ferry Van 889203	Stores van	EH
–	BR Open First 3186	Instruction Coach	DY
–	BR Open First 3381	Instruction Coach	HE
–	BR Open Standard 5636	Instruction Coach	PM
–	BR BV 6360	Barrier vehicle	NL
–	BR BV 6396	Stores van	MA
–	BR RFKB 10256	Instruction Coach	Yoker
–	BR RFKB 10260	Instruction Coach	Yoker
–	BR BFK 17156	Instruction Coach	DY
–	BR BG 92901	Stores van	WB
–	BR NL 94003	Stores van	OO
–	BR NL 94006	Stores van	OO
–	BR NK 94121	Stores van	TO
–	BR NB 94438	Stores van	TO
–	BR GUV 96139	Stores van	MA
–	BR Ferry Van 889200	Stores van	SL
–	BR Ferry Van 889202	Stores van	SL
–	BR Open Standard 975403	Cinema Coach	PM

Abbreviations:
BFK = Corridor Brake First
BG = Gangwayed Brake Van
BV = Barrier Vehicle
CCT = Covered Carriage Truck (a 4-wheeled van similar to a GUV)
GUV = General Utility Van
NB = High Security Brake Van (converted from BG)
NK = High Security General Utility Van
NL = Newspaper Van (converted from a GUV)
NW = Bullion Van (converted from a Corridor Brake Standard)
RFKB = Kitchen Buffet First

9. COACHING STOCK AWAITING DISPOSAL

This list contains the last known locations of coaching stock awaiting disposal. The definition of which vehicles are "awaiting disposal" is somewhat vague, but generally speaking these are vehicles of types not now in normal service, those not expected to see further use or vehicles which have been damaged by fire, vandalism or collision.

1252	SH	5487	CS	10588	ZH	94166	BS
1253	SH	5491	CS	10647	LM	94170	BO
1258	CS	5569	CS	10681	LM	94176	BS
1644	CS	5737	CS	10682	LM	94177	TO
1650	CS	5740	CS	10701	LM	94192	CS
1652	CS	5756	CS	10709	LM	94195	BS
1655	CS	5815	SH	10710	LM	94196	BS
1658	ZN	5876	SH	10727	LM	94197	BS
1663	CS	5888	SH	10731	LM	94199	BT
1670	CS	5900	CS	11005	LM	94207	TO
1679	ZN	5903	CS	12096	ZN	94208	TO
1680	EH	5925	SH	13306	CS	94214	CS
1696	ZN	5943	SH	13323	CS	94222	CS
1800	CS	5958	SH	13508	BO	94225	CS
2127	CS	5978	SH	17168	CS	94227	Tees Yard
2833	CS	6009	SH	18767	SH	94229	BO
2834	EH	6029	SH	18808	SH	94302	Hellifield
3241	CS	6041	CS	18862	SH	94303	Hellifield
3255	FA	6045	SH	34525	CS	94304	MH
3303	TO	6050	CS	40729	NL	94306	Hellifield
3309	CS	6073	SH	41043	LB	94308	CS
3368	FA	6134	SH	80212	CS	94310	WE
3408	CS	6151	SH	80403	CS	94311	WE
3416	CS	6154	SH	80404	CS	94313	WE
4860	CS	6175	SH	84519	CD	94316	TO
4932	CS	6179	CS	92114	ZA	94317	TO
4997	CS	6321	BO	92159	CS	94322	CS
5148	TM	6322	BO	92303	CD	94323	Hellifield
5179	TM	6323	BO	92400	CD	94326	Hellifield
5183	TM	6324	CP	92908	CS	94332	CS
5186	TM	6325	BO	92936	CD	94333	Hellifield
5193	TM	6361	NL	93723	BY	94335	BO
5194	TM	6364	WH	94101	CS	94336	BO
5221	TM	6365	WH	94104	TO	94337	WE
5331	FA	9489	CS	94106	BO	94338	WE
5386	FA	10201	LM	94116	BO	94344	TO
5420	TM	10245	CS	94137	CS	94401	CS
5453	CS	10530	ZH	94147	CS	94406	CS
5463	CS	10540	LM	94153	WE	94408	CS
5478	CS	10554	LM	94160	BT	94410	WE

94420	CS	94515	**	95763	BS	889400	ZA
94422	TO	94517	CD	96110	CS	975081	ZA
94423	BS	94520	BO	96132	CS	975280	ZA
94427	WE	94522	BO	96135	CS	975454	TO
94428	CS	94525	Hellifield	96164	CS	975484	CS
94429	Tees Yard	94526	Hellifield	96165	CS	975639	CS
94431	CS	94527	Hellifield	96170	CS	975681	Portobello
94433	BO	94528	CS	96178	CS	975682	Portobello
94434	BO	94530	CS	96182	CS	975685	Portobello
94435	TO	94531	BO	96191	CS	975686	Portobello
94445	WE	94538	CD	96192	CS	975687	Portobello
94450	WE	94539	CS	96371	WB	975688	Portobello
94451	WE	94540	TJ	96372	LM	975918	ME
94462	CD	94542	Hellifield	96373	LM	975920	Portobello
94470	TO	94545	Tees Yard	96374	ZB	977077	**
94479	TO	94546	Hellifield	96375	LM	977085	BO
94482	CS	94547	CS	96602	RU	977095	CS
94488	CD	94548	CS	96603	CF	977111	**
94490	BO	95300	CS	96604	CF	977112	**
94492	WE	95301	CS	96605	RU	977618	BY
94495	Hellifield	95400	BO	96606	RU	977989	ZA
94498	CS	95410	CS	96607	RU	999508	ZA
94501	TO	95727	WE	96608	RU		
94504	Hellifield	95754	CS	96609	RU	DS 70220	**
94512	CS	95761	WE	99019	ZR		

** Other locations:

94515	Eastleigh East Yard
977077	Ripple Lane Yard
977111	Ripple Lane Yard
977112	Ripple Lane Yard
DS 70220	Western Trading Estate Siding, North Acton

10. CODES

10.1. LIVERY CODES

The colour of the lower half of the bodyside is stated first. Minor variations to these liveries are ignored.

1	"One" (metallic grey with a broad black bodyside stripe. White National Express/Abellio Greater Anglia "interim" stripe as branding).
AG	Arlington Fleet Services (green).
AL	Advertising/promotional livery (see class heading for details).
AR	Anglia Railways (turquoise blue with a white stripe).
AV	Arriva Trains (turquoise blue with white doors).
AW	Arriva Trains Wales/Welsh Government sponsored dark & light blue.
B	BR blue.
BG	BR blue & grey lined out in white.
BP	Blue Pullman ("Nanking" blue & white).
CC	BR Carmine & Cream.
CH	BR Western Region/GWR chocolate & cream lined out in gold.
CM	Chiltern Mainline (two-tone grey & silver with blue stripes).
DR	Direct Rail Services (dark blue with light blue or dark grey roof).
DS	Revised Direct Rail Services (dark blue, light blue & green. "Compass" logo).
EC	East Coast (silver or grey with a purple stripe).
FB	First Group dark blue.
FD	First Great Western "Dynamic Lines" (dark blue with thin multi-coloured lines on lower bodyside).
FP	Old First Great Western (green & ivory with thin green & broad gold stripes).
FS	First Group (indigo blue with pink & white stripes).
G	BR Southern Region/SR green.
GA	Abellio Greater Anglia (white with red doors & black window surrounds).
GC	Grand Central (black with an orange stripe).
GN	Great North Eastern Railway {modified} (dark blue with a white stripe).
HB	HSBC Rail (Oxford blue & white).
IC	BR InterCity (light grey/red stripe/white stripe/dark grey).
M	BR maroon (maroon lined out in straw & black).
NC	National Express white (white with blue doors).
NR	Network Rail (blue with a red stripe).
NX	National Express (white with grey ends).
O	Non-standard (see class heading for details).
P	Porterbrook Leasing Company (purple & grey or white).
PB	Porterbrook Leasing Company blue.
PC	Pullman Car Company (umber & cream with gold lettering lined out in gold).
RB	Riviera Trains Oxford blue.
RP	Royal Train (claret, lined out in red & black).
RV	Riviera Trains Great Briton (Oxford blue & cream, lined out in gold).

ST	Stagecoach {long-distance stock} (white & dark blue with dark blue window surrounds and red & orange swishes at unit ends).
V	Virgin Trains (red with black doors extending into bodysides, three white lower bodyside stripes).
VN	Belmond Northern Belle (crimson lake & cream lined out in gold).
VT	Virgin Trains silver (silver, with black window surrounds, white cantrail stripe and red roof. Red swept down at unit ends).
XC	CrossCountry (two tone silver with deep crimson ends & pink doors).
Y	Network Rail yellow.

10.2. OWNER CODES

The following codes are used to define the ownership details of the rolling stock listed in this book. Codes shown indicate either the legal owner or "responsible custodian" of each vehicle.

62	The Princess Royal Class Locomotive Trust
A	Angel Trains
AV	Arriva UK Trains
BA	British American Railway Services
BE	Belmond (UK)
BK	The Scottish Railway Preservation Society
DB	DB Schenker Rail (UK)
DR	Direct Rail Services
E	Eversholt Rail (UK)
EM	East Midlands Trains
EU	Eurostar (UK)
FG	First Group
LS	Locomotive Services
NR	Network Rail
NY	North Yorkshire Moors Railway Enterprises
P	Porterbrook Leasing Company
RA	Railfilms
RP	Rampart Engineering
RV	Riviera Trains
VT	Vintage Trains
WC	West Coast Railway Company

10.3. OPERATOR CODES

The two letter operator codes give the current operator (by trading name). This is the organisation which facilitates the use of the coach and may not be the actual Train Operating Company which runs the train on which the particular coach is used. If no operator code is shown then the vehicle is not in use at present.

62	The Princess Royal Class Locomotive Trust
AW	Arriva Trains Wales
BK	The Scottish Railway Preservation Society
BP	Bemond British Pullman

CR Chiltern Railways
CS Colas Rail
DB DB Schenker
DR Direct Rail Services
EC East Coast
EM East Midlands Trains
EU Eurostar (UK)
GA Abellio Greater Anglia
GB GB Railfreight
GC Grand Central
GW First Great Western
NB Belmond Northern Belle
NY North Yorkshire Moors Railway
RP Royal Train
RS The Royal Scotsman (Belmond)
RV Riviera Trains
SR ScotRail
ST Statesman Rail
VT Vintage Trains
VW Virgin Trains
WC West Coast Railway Company
XC CrossCountry

10.4. ALLOCATION & LOCATION CODES

Code Depot   *Operator*

AL	Aylesbury	Chiltern Railways
BH	Barrow Hill (Chesterfield)	Barrow Hill Engine Shed Society
BN	Bounds Green (London)	East Coast
BO	Burton-upon-Trent	Nemesis Rail
BQ	Bury (Greater Manchester)	East Lancashire Rly Trust/Riley & Son (Railways)
BS	Bescot (Walsall)	DB Schenker Rail (UK)
BT	Bo'ness (West Lothian)	The Bo'ness & Kinneil Railway
BY	Bletchley	*Storage location only*
CD	Crewe Down Holding Sidings	Riviera Trains
CL*	Crewe LNWR Heritage	London & North Western Railway Company
CF	Cardiff Canton	Arriva Trains Wales/Colas Rail
CM	East Cranmore	Cranmore Railway Company
CP	Crewe Carriage Shed	LNWR Company (part of Arriva)
CS	Carnforth	West Coast Railway Company
DY	Derby Etches Park	East Midlands Trains
EC	Edinburgh Craigentinny	East Coast
EH	Eastleigh	LNWR Company (part of Arriva)
FA	Fawley (Hampshire)	*Storage location only*
HE	Hornsey (London)	Govia Thameslink Railway
IL	Ilford (London)	Abellio Greater Anglia
IS	Inverness	ScotRail
KM	Carlisle Kingmoor	Direct Rail Services

KY	Knottingley	DB Schenker Rail (UK)
LA	Laira (Plymouth)	First Great Western
LB	Loughborough Works	Wabtec Rail
LM	Long Marston (Warwickshire)	Motorail (UK)
MA	Alstom Longsight (Manchester)	Alstom
ME	Mossend Yard	DB Schenker Rail (UK)
MH	Millerhill (Edinburgh)	DB Schenker Rail (UK)
NC	Norwich Crown Point	Abellio Greater Anglia
NL	Neville Hill (Leeds)	East Midlands Trains/Northern
NY	Grosmont (North Yorkshire)	North Yorkshire Moors Railway Enterprises
OO	Old Oak Common HST	First Great Western
PG	Peterborough	GB Railfreight
PM	St Philip's Marsh (Bristol)	First Great Western
PZ	Penzance Long Rock	First Great Western
RU	Rugby	Colas Rail
SH	Southall (London)	West Coast Railway Co/Locomotive Services
SK	Swanwick Junction (Derbyshire)	Midland Railway Enterprises
SL	Stewarts Lane (London)	Southern/Belmond
SP	Springs Branch (Wigan)	DB Schenker Rail (UK)
TJ	Tavistock Junction Yard (Plymouth)	*Storage location only*
TI	Temple Mills (London)	Eurostar (UK)
TM	Tyseley Locomotive Works	Birmingham Railway Museum
TN	Thornton (Fife)	John Cameron
TO	Toton (Nottinghamshire)	DB Schenker Rail (UK)
TW*	Tonbridge West Yard	GB Railfreight
WB	Wembley (London)	Alstom
WD	Wimbledon (London)	South West Trains
WE	Willesden Brent Sidings	*Storage location only*
WH	Washwood Heath (Birmingham)	Boden Rail Engineering
YK	National Railway Museum (York)	National Museum of Science & Industry
ZA	RTC Business Park (Derby)	Railway Vehicle Engineering
ZB	Doncaster Works	Wabtec Rail
ZC	Crewe Works	Bombardier Transportation UK
ZD	Derby Works	Bombardier Transportation UK
ZG	Eastleigh Works	Arlington Fleet Services
ZH	Springburn Depot (Glasgow)	Knorr-Bremse Rail Systems (UK)
ZI	Ilford Works	Bombardier Transportation UK
ZJ	Stoke-on-Trent Works	Axiom Rail (Stoke)
ZK	Kilmarnock Works	Wabtec Rail Scotland
ZN	Wolverton Works	Knorr-Bremse Rail Systems (UK)
ZR	York (Holgate Works)	Network Rail

*= unofficial code.